TRANSACTIONS OF THE

AMERICAN PHILOSOPHICAL SOCIETY

HELD AT PHILADELPHIA

FOR PROMOTING USEFUL KNOWLEDGE

———

VOLUME 69, PART 6 · 1979

Henrik Krøyer's Publications on Pelagic Marine Copepoda (1838–1849)

CARL C. DAMKAER

AND

DAVID M. DAMKAER

DEPARTMENT OF OCEANOGRAPHY, UNIVERSITY OF WASHINGTON

THE AMERICAN PHILOSOPHICAL SOCIETY

INDEPENDENCE SQUARE: PHILADELPHIA

September 1979

Library of Congress Catalog
Card Number 79–51538
International Standard Book Number 0–87169–696–7
US ISSN 0065–9746

HENRIK KRØYER'S PUBLICATIONS ON PELAGIC MARINE COPEPODA
(1838–1849)

Translated and Edited by CARL C. DAMKAER and DAVID M. DAMKAER*

CONTENTS

INTRODUCTION

Henrik Krøyer pioneered in the study of Copepoda during the earliest years when oceanic species were being discovered and described by a number of outstanding workers, notably James Dwight Dana and Henri Milne-Edwards. From a mid-twentieth-century position, it is difficult to appreciate the new ground gained by Krøyer and his contemporaries. Published works on copepods before 1840 were few and often obscure, and, though there were some exceptions, most were woefully inadequate.

Few present-day workers have access to original Krøyer publications, all of which are rare. Fortunately, Krøyer's work was given due regard and much was incorporated into reviews published in other languages, particularly German and English. Recent interest in the systematic position of the species of *Calanus, Pseudocalanus,* and other genera, some species of which were described first by Krøyer, has increased the importance of this early work and made accessibility mandatory. Beyond that, there is considerable historical value in Krøyer's papers. Therefore, we have translated Krøyer's works on free-living copepods, to make them available, with his illustrations, in a single document.

* Contribution No. 1030 from the Department of Oceanography, University of Washington WB-10; Seattle, Washington 98195.

Today, Krøyer's Danish makes difficult reading because of the archaic style and frequent use of long and complex sentences. It was not our intention to rewrite Krøyer's papers, so we have retained as much of his original style as possible, consistent with clarity in English. Editorial comments have been placed within brackets [ ]; parentheses ( ) are Krøyer's own. In addition to the Danish text, Krøyer included succinct Latin diagnoses for each species. The language of the various sections is indicated. Krøyer began the 1848–1849 series with "I. Calanus," but he did not continue this Roman numbering sequence. Species were assigned Arabic numerals, and we have added these, in brackets, where Krøyer inconsistently omitted them.

Most of the morphological terms used by Krøyer are unambiguous, though quaint by today's standards. Krøyer believed, with many contemporaries, that the second maxillae and the maxillipeds formed a functional and anatomical group, and referred to them as "first pair of jaw-feet" and "second pair of jaw-feet." He also believed that the first maxillae were two pairs of appendages, although he acknowledged that if so they had a common base and that he was uncertain in assigning the various parts. His collective term for the first maxilla is "jaws," while he referred to the gnathobase as the "actual jaw" and to various inner lobes (including the gnathobase) and outer lobes as "first jaw-pair" and "second jaw-pair," or sometimes "jaw-plates." Where Krøyer used the collective term, we use the modern equivalent, first maxillae. Otherwise the literal translation is retained. Krøyer's term for mandible translates directly.

Krøyer's measurements are given in "lines," which we have interpreted as 1/12 British inch, or 2.12 mm, and converted to metric units. Copepod lengths were measured along the dorsal midline from the anterior edge of the head to the end of the caudal rami, but not including the caudal setae. (Dr. Torben Wolff has kindly informed us that Krøyer's lines could be based on the old Danish inch [1 line = 2.18 mm], so that all measurements may be increased by three per cent.)

A substantial part of Krøyer's work on pelagic copepods first appeared in the *Atlas de Zoologie*, without date. When it was apparent to Krøyer that no text would accompany the plates, he published his manuscript in the *Naturhistorisk Tidsskrift* (1848–1849). It may not now be possible to determine the date of publication of the *Atlas*. Nomenclatural problems involving the genera first mentioned there have been resolved, and would not be altered even if priority could be proved (6).[1] Since Krøyer cited the *Atlas* in 1847, it probably pre-dates that year (8). Neave (13) cited the date of the *Atlas* as 1842, but it is not likely that the entire *Atlas* was published then. With (16) was the academic descendant of Krøyer, and he examined and redescribed several of Krøyer's type-specimens. With gave the *Atlas* date as 1842–1845, the dates also given earlier by Giesbrecht (5). Possibly the plates were published over several years. In the present work we have somewhat arbitrarily indicated the date of the *Atlas* as [1845].

We are indebted to Cornell University Library for negatives of *Atlas* plates 41–43, reduced and reproduced (without color) in the present work.

HENRIK NIKOLAI KRØYER
(1799–1870)

LIFE

Henrik Krøyer was born 22 March, 1799, in Copenhagen, of a family of seafarers, although his father was a bookkeeper in the Office of Pensions. Krøyer's earliest schooling leaned toward history and classics, which had a lifelong influence on his writings. He had inherited his grandfather's books, which impressed him greatly, and with which he spent much time reading in many languages.

Krøyer's father was transferred to the West Indies about 1813, and died there about 1817. It had been arranged for Krøyer to live in the home of a prominent physician, F. L. Bang, and Krøyer became a university student in medicine. He retained a strong interest in history and philology and was also active in organized student affairs.

Youthful idealism overtook Krøyer in 1821, when he sided with the Greeks against "Turkish oppression." Krøyer joined other young men who would fight for the Greek cause alongside the descendants of ancient Greece. With many adventures, Krøyer traveled by foot across Europe to Marseille, where the "friends of the Greeks" were detained for some months. Rather than join in the loafing and carousing, Krøyer studied Greek classics in the original language, but was also attracted to sea life examined firsthand or by way of fishermen's catches.

After a perilous crossing to Greece, Krøyer found that the Greece of his day did not resemble his ideals of ancient Greece, and he was overwhelmingly disappointed. Their poor treatment caused Krøyer and

---

[1] Numbers in parentheses indicate references listed on page 9.

Henrik Nikolai Krøyer (1799–1870) (from Dahl, 1941*b*).

others of the group to return, and Krøyer nearly reversed his sympathies in this conflict.

With much difficulty, Krøyer reached Rome, and eventually Germany, where he remained for a year at the Universities of Heidelberg and Göttingen. Krøyer attended lectures and read widely, and began a formal interest in natural history. In 1823 he returned on foot to Copenhagen, as he stated, poor in illusions and money, but rich in experiences and with his humor intact. Krøyer continued his studies, primarily in classics, but he did not take a degree.

Krøyer had difficulty finding secure instructional employment. In 1827 he obtained a teaching position in a Latin school in Stavanger [Norway], where he remained for three years, continuing his interest in historical and philological subjects. Krøyer prepared a topographic work on Stavanger, which led him into the natural history of the region. The plants had been adequately covered by the *Flora Danica*, but Krøyer wished to describe the animals. To that end he gathered and maintained a small menagerie, especially of birds, and he obtained some unusual fish and other marine organisms from fishermen. Little or nothing resulted from these efforts, but Krøyer now had a serious introduction to natural history, and his studies took a new direction (2, 3, 4).

In 1830 Krøyer became seriously ill, possibly with meningitis, and wavered between life and death for six months. In his own words, Krøyer had no confidence in his doctor, a German who had been a barber, and still less in the doctor's assistant, a bookbinder boy. Krøyer decided to return to Copenhagen. A young girl who had been caring for him in his home accompanied him on the trip. This girl, Bertha Cecilie Gjesdal (born 1817), returned to Stavanger, but in 1833 became Krøyer's wife (12).

Krøyer recovered and began to study natural history at several schools in Copenhagen. He also sought a teaching position, but now his field was natural history; he obtained part-time instructor's posts in at least two schools in Copenhagen by 1833.

The exposure to the fish and fisheries of Marseille and Stavanger resulted in Krøyer's dissatisfaction with only teaching. Krøyer conceived a description of the whole Danish fisheries, including the natural history of the fishes and how they were caught in the various parts of the country. For support, Krøyer turned to the Ministry of Agriculture and its leader Jonas Collin, a well-known benefactor of natural sciences. For 1834 and 1835, Collin arranged to grant Krøyer a considerable sum, compared to his previous financial situation, and Krøyer abandoned his school connections.

In 1834 Krøyer and his young wife sailed in an open boat around the Danish coast, obtaining unique and first-hand knowledge of fish populations and the fisheries. Krøyer learned much, for he had an ability to gain confidence and speak effectively to the fishermen. In the first year most of the east coast and some of the west coast of Jutland was surveyed. In 1835 Krøyer visited the rest of Jutland's east coast, the islands, and the south coast of Sweden (2, 11).

Krøyer's modest voyage continued a tradition begun the previous century by the Danish naturalists Otto Fabricius, Peter Forsskål, and Otto Friedrich Müller. These remarkable men were strong individualists, whose lives developed in very different ways, yet they had in common an inspiration to add to knowledge of marine life. And each had a competence which made their life work outstanding.

Danish natural science did not prosper during the early 1800's. This was in part due to economic factors, the country becoming utterly bankrupt in 1813. In 1824 H. C. Ørsted founded the Society for the Propagation of Natural Science, heralding increased interest and activity in the 1830's. It was at this favorable time that Krøyer became totally committed to zoology. A group of influential men, including Krøyer, formed the Danish Natural History Association in 1833. The Association promoted lectures and sponsored a small museum to foster wider appreciation of natural history. There was a Royal Museum of Natural History, but it was not open to the public, and the Association, as well as the earlier Society, tried without success to alter this policy. In 1835 the Association's museum received a large collection from Krøyer, derived from his investigations of the Danish fisheries. This collection ultimately became an important part of the Royal Museum's extensive Danish fauna series (2, 17).

In 1836 Krøyer made a significant and timely contribution to the development of Danish natural history by founding and editing the Naturhistorisk Tidsskrift. To aid this new journal, the government subscribed to seventy-five copies for distribution to libraries and universities, although some of the expenses were borne by Krøyer himself. In a short time, Krøyer and other contributors, with outstanding scientific reports, built a strong reputation for the Tidsskrift. It was seriously referred to abroad in spite of being published in the Danish language; some articles were translated into German and French and widely circulated. J. C. Schiødte became editor of the Tidsskrift after Krøyer's death, and this journal continued until Schiødte died in 1884 (2, 11).

Inevitably, Krøyer's name became well known outside of Denmark, and in 1838 he was invited by the French government to participate in an expedition to Spitsbergen on the corvette La Recherche. Krøyer met the expedition in Trondheim in June, investigating northern Norway for a few weeks before heading toward Spitsbergen. The vessel remained only eleven days at Spitsbergen, returning to Norway in mid-August. Until mid-September Krøyer spent a few days collecting at each of several points from North Cape to Tromsø. Krøyer divided the winter between Tromsø and Lofoten, where his attention was given to the fisheries (3).

The voyage by Krøyer and the other members of this expedition was among the earliest marine investigations in that region. In the two hundred years after Dutch navigators recorded Spitsbergen in 1596, it was visited by a succession of English, Norwegians, Russians, Basques, and Dutch. They were attracted particularly by seals and whales, and also because Spitsbergen was a landmark for further Arctic exploration. The first scientific expedition seems to have been that of the Swedish geologist B. M. Keilhau in 1827. Another Swede, Sven Lovén, dredged along the Spitsbergen coast in 1837. He was followed by La Recherche the next year, and then by a series of scientific expeditions, mainly from Norway and Sweden (14).

Krøyer felt that the expedition was poorly organized, and that its leader, Paul Gaimard, lacked energy and understanding, and in general was immature. Also Krøyer said that La Recherche was too small for so many participants, which included naturalists from Sweden and Norway besides France and Denmark. Apparently, the expedition had promised Krøyer more than it finally delivered. The plans included a stop at Bear Island, but this important part of the expedition was excluded. And the plans called for a four- to five-week stay in Spitsbergen, but, in Krøyer's words, "a few pieces of ice showed up at the entrance to the fjord on July 30, and started a panic, and that had something to do with us leaving early." In spite of these and other frustrations, Krøyer said that he had interesting results and was satisfied because of this unique collecting opportunity. Overall good weather made the trip pleasant (3).

Krøyer and some of the other expedition members remained in the north at one task or another to participate in a planned second Spitsbergen trip in the following year. However, when Krøyer believed in April, 1839, that such would not materialize, he returned directly to Bergen and then to Copenhagen. Some participants had already returned to France; those others that had stayed the winter, including Gaimard, went through Lapland to Stockholm. Krøyer said that this overland phase of the expedition would not have had zoological interest for him (3).

The results of this expedition are poorly known because of the failure to publish a text of the reports and because of the rarity of the large, splendid *Atlases*, of which there were at least two, both undated: *Atlas Géologique* and *Atlas de Zoologie*.

In late summer 1840 Krøyer accompanied the Danish navy frigate *Bellona* to South America. For the government, this one-year voyage was to establish diplomatic contacts in several countries and to negotiate commercial treaties. For Krøyer, this was again an opportunity to make important collections, particularly near Madeira, Brazil, Chile, and in the open ocean. Krøyer used a "bunting bag" plankton net in some of his sampling, and presumably he had also done so on *La Recherche* (9, 17).

In 1842 Krøyer became an unpaid assistant, and subsequently an equally unpaid curator in zoology, at the Royal Museum of Natural History. However, in 1845 with the death of the chief curator of zoology, J. Chr. H. Reinhardt, that position, as well as the corollary appointment as professor of zoology at the University of Copenhagen, went to the younger Japetus Steenstrup (1813–1897), and not to Krøyer as he expected. Krøyer was deeply hurt and attempted to resign his position, but could not obtain the required permission. The king personally told Krøyer not to leave the museum, and that if Krøyer needed more money he should teach. Krøyer, in his words, began to consider himself one of the museum's living exhibits (2, 3).

Undoubtedly, Krøyer had a somewhat trying personality which may have kept the desired appointment from him (2). Michael Sars, meeting him in 1837,

considered him "most valuable" though "not really good-natured"; in 1856 Sars recommended Krøyer as one of the most knowledgable naturalists of the northern countries (18). And, too, Krøyer had a non-traditional family life which probably also caused tongues to wag (12).

Time has shown that Steenstrup was also an extraordinary naturalist, but from this experience came a lifelong fight between Krøyer and Steenstrup, which placed its mark on many Danish zoologists. Krøyer was made inspector of museum, a salaried administrative position, but from this time onward, he was very bitter and despondent, and he increasingly withdrew to himself (2, 3).

Still despondent, in 1853 Krøyer made a long trip to the coasts of Europe and North America. From this came large collections from Scotland, Ireland, Holland, France, Newfoundland, and the Gulf of Mexico. In his collection and inspection of fishes, Krøyer never neglected to look for parasitic Crustacea, and in this way obtained much material that until then had been essentially unknown. His published specimen locality records give scattered hints of his itinerary: Baltimore; "a longer stay in Biloxi" [Mississippi]; New Orleans; "near St. Louis"; Cincinnati (9, 11, 17). The indefatigable Danish naturalist caused enough of a stir for S. F. Baird to send an alarm to Louis Agassiz (1 June, 1854):

I have just learned that Henry Kroyer of Copenhagen has been spending a month or two about New Orleans making immense collections of fishes and other things and is now proceeding up the Mississippi. He has been paying particular attention to the cyprinodonts and he has told my informant that he had collected 6 or 8 new species in a single day; also that he had discovered great differences between ♂ and ♀ and all that! He must be back in Copenhagen in August when he will doubtless publish his collection of new ones. Now why dont you complete your manuscript of cyprinodonts and publish in the Journal of Science the characters at once, giving diagnosis of genera and species? I should feel very much grieved to see your species and mine taken away thus when we were the first to occupy the field.

Agassiz thanked Baird but declined to hurry his descriptions (7). It is clear that Krøyer was only able to assimilate a small part of these collections for his own research.

Eventually, ill health accompanied Krøyer's disappointment. In his later years, from about 1867, he often was unable to go to the museum. By 1869, though his mental powers never failed him, he was almost an invalid, and reluctantly resigned his position. Krøyer died on 14 February, 1870 (2, 4).

Krøyer was isolated by his imagination and bitterness, but ultimately also by poor health. Nevertheless, there were many who respected and honored him for his substantial contributions to zoology. He had been awarded an honorary doctorate from the University of Rostock [Germany] (3).

SCIENTIFIC WORK

Krøyer's largest contribution was his work on Denmark's fishes. This began with his explorations along the Danish coast in 1834. His *Danmarks Fiske* was published in booklet form from 1838, and by 1853 the work comprised three large volumes. *Danmarks Fiske* was supported in part by the Ministry of Agriculture, especially Jonas Collin (to whom Krøyer dedicated one of the volumes), and by royal grants (2).

Krøyer's financial situation remained precarious. He was able to sell the separate issues of *Danmarks Fiske* for profit. He received limited support for his coincident publication, *Naturhistorisk Tidsskrift*. Each year there were at least three issues of the *Tidsskrift*, of which Krøyer had to buy seventy copies with one-half of his government grant. More issues in a year used more of the grant for publication costs, and Krøyer's share became, in his own words, in opposite proportion to his work (3).

For *Danmarks Fiske*, Krøyer intended to examine a large number of specimens of each species, considering the morphology of each sex and age from different places and seasons. On this there had been very little knowledge before Krøyer, who revised and increased this many times over. Krøyer also discussed each species's way of life and world distribution. He planned to include economic and technical sections, but only brief comments were published. There remain in manuscript enormous quantities of material on the early Danish fisheries (2, 17).

RESEARCH ON CRUSTACEA

Krøyer published a great number of short papers on Crustacea, mostly in the *Naturhistorisk Tidsskrift*. These were bound collectively and distributed by Krøyer as his *Opuscula carcinologia* [1836–1849]. Dealing primarily with morphology and systematics, these papers won Krøyer a leading position in this field, and are considered classics within zoology. Subjects included free-living and parasitic copepods, barnacles, amphipods, isopods, tanaids, mysids, cumaceans, crabs, and shrimps. Krøyer also wrote about pycnogonids and their development, mollusks, echinoderms, whales, and fossil mammals. Krøyer had access to his own collections, which were considerable and far beyond those of any of his countrymen. In addition, Krøyer received the cooperation of several highly reliable travelers, including ship-captains, government officials, and scientists. They brought Krøyer specimens from Greenland, Iceland, Central America, the West and East Indies, Kamchatka, and elsewhere. Krøyer gratefully acknowledges this assistance in his various papers.

Krøyer's first-hand experiences in both northern and tropical regions, together with the material he received from his associates, made him appreciate the variety of high latitude animal life. Krøyer accepted the widely held "law" that there were many more marine species and families in the tropics than in polar zones. But Krøyer demonstrated such an abundance of cold-water species that he dispelled the belief that the northern seas were altogether unfavorable to variation in animal forms.

Krøyer's largest single work on Crustacea was his [Contributions to knowledge of parasitic Crustaceans] (1863) (9). Most of Krøyer's papers were characterized by a monographic approach, thoroughly reviewing families or genera on a systematic basis, rather than by considering many groups from a single region. Krøyer's illustrations and tables of measurements provided much more detail on the organisms, their segmentation and their appendages, than that given by most of his contemporaries.

KRØYER'S COPEPODS

Krøyer's studies on copepods added or expanded several fruitful lines of investigation. His descriptions routinely considered details beyond most contemporary work. Krøyer cautioned his readers on morphological variability, because of which he attempted to observe and measure many specimens; he stressed what he found to be constant proportions and measurements. Besides morphology, Krøyer recorded geographical distribution, color, and behavior, and sometimes internal anatomy and responses of organs in living specimens. Krøyer mentioned the esthetes on the first antennae of male *Calanus*, the constant feeding or respiratory movement of *Calanus* mouthparts, parasites on copepods, and associated planktonic organisms. Krøyer commented on the large quantity of oil in *Calanus*, relating this to its use as whale food; he speculated on the potential economic value in obtaining this oil directly.

Krøyer was the first to place correctly Gunnerus's *Monoculus finmarchicus* [1770] in the genus *Calanus* Leach, 1819, and to place *Cetochilus* Roussel de Vauzeme, 1834 in synonymy with *Calanus* (10). The principal reviewer after this time, Claus [1863], did not have access to Krøyer's papers (1), so it was not until Giesbrecht's [1892] great monograph that Krøyer's analysis was given general recognition (5).

Krøyer sought without success to find asymmetry in first antennae and fifth legs, and therefore sexual differences, in *Calanus*, such as were known in other genera, e.g. *Anomalocera* and *Candacia*. Nor was he aware that the number of urosomal segments was not a systematic character. Therefore, Krøyer did not recognize sexual differences within *Calanus*, so that he described the male and female, and also the stage V juvenile, as separate species. Krøyer humbly stated that further workers might well synonymize some of these species. Again ahead of his time, Krøyer deposited his specimens in a museum, which was specifically cited in the published report.

It had been recognized since the work of Jurine [1820], that the nauplius was merely a stage in the life cycle of these crustaceans. However, Krøyer was the first to describe and illustrate the nearly complete development of a marine calanoid copepod (10). He recognized that the nauplius and metanauplius were part of this cycle, although he believed that copepodid IV was the oldest larva, and that, for *Calanus hyperboreus*, copepodid V was the adult [rather than copepodid VI]. Krøyer realized that specific or even generic characters did not appear in the early stages, and that he possibly dealt with more than one species in this cycle. He explained that this indeed was likely because of the great size differences within each stage, but that his comments would hold for the development of *Calanus* in general. [*Calanus finmarchicus, helgolandicus, glacialis,* and *hyperboreus* are all found in Krøyer's research area.]

In his work on the parasitic copepods, Krøyer was the first to point out that *Chalimus scombri* Burmeister, 1835 was probably a juvenile stage of a caligid.

KRØYER'S ILLUSTRATIONS

The illustrations in Krøyer's publications deserve separate mention because they introduce an extraordinary aspect of his life. It is not always clear which drawings were made by Krøyer himself, although most are signed by him or others under his direction. The 1838 figure of *Calanus hyperboreus* suggests a medieval woodcut, and certainly Krøyer was capable of better. At that time Krøyer had only one specimen which he did not want to dissect; he later [1848] apologized for the imperfect description. The illustrations of amphipods in this same early paper are very good. Several of Krøyer's early carcinological papers in the *Tidsskrift* have excellent figures drawn by his student Jørgen Christian Schiødte (1815–1884), who enjoyed a lifelong reputation for such work (11). Other illustrations, also of high quality, are signed by Krøyer.

The copepod plates in *Atlas de Zoologie* were drawn by one C. Thornam. Some of these are similar to the 1849 *Tidsskrift* plate by Schiødte, but it is not possible to determine which was drawn first. The *Atlas* figures are the better, and were printed in subtle and beautiful colors.

The eighteen engraved plates in Krøyer's 1863 monograph on parasitic copepods (9) are excellent, and the fact that they were drawn by a nine-year-old boy is remarkable. This boy was born to Krøyer's mentally retarded sister-in-law, Ellen Cecilie Gjesdal, in Stavanger in 1851. The father's identity, though there was much subsequent conjecture, was never proved. The child, Peder Severin Krøyer (1851–1909), was placed in the care of Krøyer's wife in Copenhagen, and Krøyer became his foster father. Peder Krøyer drew from the time he could hold a

Henrik Krøyer, drawn by his foster son P. S. Krøyer *ca.* 1868 (from Fiedler, 1888).

pencil. When he was nine, Peder was called one evening to his foster father's study. Krøyer was looking through a microscope at parasitic copepods and he asked Peder whether he could draw them. Peder saw for the first time "a whole new world of wonderful small creatures," and draw them he did. When the drawings were finished, Krøyer used them in his monograph. P. S. Krøyer came to be one of Denmark's most well-known artists, and a leader of a naturalistic school which flourished around 1900 (12).

Krøyer lived in Strandgade on Christianshavn in one of Copenhagen's oldest houses. The house had belonged to the town's burgomaster, and had been used in 1619 by Jens Munk in planning a northwest passage. Krøyer lived among heavy, sturdy furniture in an atmosphere of quiet which he demanded (12). With a frugal mode of living, within his poor salary, Krøyer managed to build a library of 12,000 titles, probably the largest private zoological library in Denmark (2). Every evening he worked with his microscope and books. At times, three Gjesdal sisters lived at the Krøyer house. Some said that Krøyer's wife followed him more as a servant than as a wife; though she loved and admired Krøyer, all her life she felt inferior to him. Krøyer had a son by Inger Gjesdal, Vilhelm Krøyer (1844– ?), who became a distinguished and popular English teacher in Denmark (12).

SUMMARY

The professional jealousies and bitterness between Krøyer and Steenstrup were apparent in the publications of each. Neither let an opportunity pass to be openly critical of the other. This was especially true in their studies of parasitic copepods which totally overlapped. The strength of Krøyer and his followers lay in careful, detailed microscopic studies, and they had little esteem for Steenstrup's group. This feeling was mutual, and affected the development of zoology in Denmark for seventy-five years. Krøyer's student Schiødte inherited the combat, and his student H. J. Hansen (1855–1936) and Hansen's student Carl With (1877–1923) in turn were plagued by this schism long after both Schiødte and Steenstrup had died (15).

Henrik Krøyer was the most important Danish zoologist of the mid-nineteenth century. His name has been memorialized in the genera *Krøyera* Bate, 1855 [Amphipoda], and *Krøyeria* van Beneden, 1853 and *Krøyerina* Wilson, 1932 [Copepoda], as well as in several specific names.

LITERATURE CITED

1. Claus, C. 1863. *Die frei lebenden Copepoden mit Berucksichtigung der Fauna Deutschlands, der Nordsee und des Mittelmeeres* (Leipzig, Engelmann).
2. Dahl, Svend. 1941a. *Den Danske Plante- og Dyreverdens Udforskning* (København, Udvalget for Folkeoplysnings Fremme, G. E. C. Gad).
3. ——. 1941b. *Til Minde om zoologen Henrik Krøyer* (København, Alfred G. Hassing).
4. Fiedler, H. V. 1888? "Om Professor Henrik Krøyer." *Nordisk Aarsskrift for Fiskeri* 1887: pp. 1–13.
5. Giesbrecht, Wilhelm. 1892. "Systematik und Faunistik der pelagischen Copepoden des Golfes von Neapel und der angrenzenden Meeres-abschnitte." *Fauna und Flora des Golfes von Neapel* 19: pp. 1–831.
6. Grice, G., and W. Vervoort. 1963. "*Candacia Dana*, 1846 (Crustacea: Copepoda): Proposed Preservation under the Plenary Powers and Designation of a Type-species for the Genus in Accordance with Common Usage." *Bull. Zoological Nomenclature* 20, 2: pp. 150–152.
7. Herber, E. C. (ed.). 1963. "Correspondence between Spencer Fullerton Baird and Louis Agassiz—Two Pioneer American Naturalists." *Smithsonian Inst. Publ.* 4515: pp. 1–237.
8. Krøyer, Henrik. 1847. "Karcinologiske Bidrag (Fortsaettelse). Beskrivelsen af nordiske Isopoder." *Naturhistorisk Tidsskrift*, new ser., 2, 4: pp. 366–446.
9. ——. 1863. "Bidrag til Kundskab om Snyltekrebsene." *Loc. cit.*, ser. 3, 2: pp. 75–426.
10. Marshall, S. M., and A. P. Orr. 1955. "The Biology of a Marine Copepod, *Calanus finmarchicus* (Gunnerus)." (Edinburgh, Oliver and Boyd).
11. Meinert, Frederik V. A. 1902. "Zoologi." *In*: P. Hansen (ed.). *Illustr. dansk Litteratur-historie* 3, 2: pp. 1049–1093.
12. Mentze, Ernst. 1969. *P. S. Krøyer, Kunstner af stort format—med braendte vinger* (København, Schønbergske Forlag).
13. Neave, S. A. (ed.). 1940. *Nomenclator Zoologicus* (4 v., London, Zoological Society of London).
14. Nordenskiöld, A. E. 1879. *The Arctic Voyages of Adolf Erik Nordenskiöld 1858–1879* (London, Macmillan).
15. Stephensen, K. 1937. "H. J. Hansen (1855–1936)." *Vidensk. Medd. fra Dansk naturh. Foren.* 100: pp. v–xii.
16. With, Carl. 1915. "Copepoda. I." *Danish Ingolf-Expedition* 3, 4: pp. 1–262.
17. Wolff, Torben. 1967. *Danske Ekspeditioner på Verdenshavene* (Copenhagen, Rhodos International Science and Art Publishers).
18. Økland, Fridthjof. 1955. *Michael Sars, et Minneskrift* (Oslo, Norske Videnskaps-Akademi [Jacob Dybwad]).

TRANSLATIONS OF HENRIK KRØYER'S PUBLICATIONS ON PELAGIC MARINE COPEPODA

(1838) "Greenland Amphipods (An addition: Description of some other crustaceans from Greenland, and an enumeration of the known Greenland species of the crustacean class, in connection with some zoological-geographical remarks about the boreal Crustacea)," *Kongelige Danske Videnskabernes Selskab Skrifter, naturvidenskabelig og mathematisk Afdeling*, series 4, 7: pp. 231–326, 4 pls.

[Footnote] The material which I have used for this present work can be found stored partly in the Royal Museum and partly in the Natural History Society's collection. Professor Reinhardt, who has allowed me freest admission to the Museum's Greenland crustaceans, had at an earlier time already assigned names to some of the new species, which names I have naturally seen as my duty to preserve. This species described here [*Lysianassa vahlii*] he named after the sender, Mr. I. Vahl (a son of the famous botanist by the same name), who during a long stay in Greenland had sent down many interesting contributions to the fauna of that country.

Description of some Greenland
crustaceans other than in the amphipod-order.

Calanus hyperboreus Kr. [new species]
(Pl. IV, fig. 23).

This crustacean, new for Greenland's fauna, is fairly transparent, and, at least after having been stored in alcohol for a long time, has a likeness to the discarded exuviate of an animal rather than of a complete animal.

The head and the body, which are closely united, together show almost the form of a half-cylinder; a thin, elongate tail or urosome extends from the body. —*The length* from the edge of the forehead to the tip of the tail is 8.5 mm, of which the tail amounts to almost 2.1 mm.

The head is large, insignificantly longer than the body's first two segments, or amounting to almost ⅓ the length from the edge of the forehead to the beginning of the tail. The edge of the forehead protrudes forward slightly in the shape of a bulge, and slopes steeply after that.

The antennae, of which I *only noticed one pair, in spite of giving them the most careful inspection,* are of a comparatively very strong construction, and distinguish themselves by a considerable length, which is just as long as the length from the edge of the forehead to the tip of the tail, or even slightly longer. They are not separated into shaft and flagellum, but consist of some twenty identical segments. A very strange peculiarity shows in the construction of the last segments: a very long feather (i.e. a seta, which sends out lateral setae) extends from the end of the fourth-last segment, beside and below the third-last segment; the third-last segment likewise sends out a long feather, but from a knot on the underside, a little ways from the end of the segment; the last two segments are very small, and the last ends with two bunches of setae, but the above-mentioned feathers reach far out over these. The antennae are supplied with setae along the anterior edge.

The eyes' pigment had completely disappeared by being stored in alcohol; yet I have observed the eyes in the form of circular points on the anterior edge of the head.

The head, on the underside, shows three pairs of organs, which in formation seem to have a likeness to feet, but in all probability are *mouthparts;* at least I have not been able to discover any other mouthparts. I am not able to give any detailed description of the character of these organs, because of defective individuals due to damage. I will only remark, that the last pair is the largest, and that it has the most conspicuous likeness to feet.

The body consists of five segments, of which the posterior one is shorter than the anterior one, yet in only a small degree.

Each of the body's segments are supplied with a *pair of swimming-feet;* these consist of two elongate segments and a five-segmented ramus, the posterior edge of which is closely set with long setae.

The tail consists of five segments, which gradually get narrower; the last one is cleft to the base, and is shaped like two somewhat oval plates, which get slightly wider toward the end where each of them is armed with five long setae. The length of these setae is almost like the length of the tail's last three segments. The tail's second segment surpasses the others in length, yet not considerably.

Mr. Vahl has sent this animal down from Godhavn.

Figures: Pl. IV, fig. 23, *a. Calanus hyperboreus;*

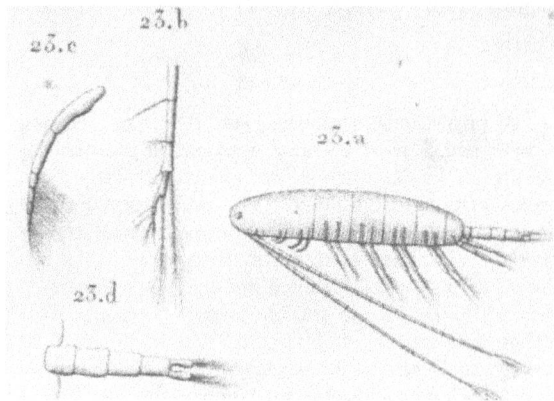

Calanus hyperboreus Krøyer, 1838.

b. last segment of antenna, greatly enlarged; *c.* a swimming-foot; *d.* the tail's segments, from above.

Of all the crustaceans known to me, this seems to come closest to *Cyclops longicornis,*[1] yet in this way, that at a glance one can easily see that they are separate species. *Leach* regarded the above-named *C. longicornis* as the type of a new genus, *Calanus,* which is characterized by having only two antennae [i.e. one pair; *Cyclops sensu strictu* was said to have four antennae, i.e. two pairs]. *Latreille,* who doubted the correctness of this condition,[2] did not dare to take up Leach's genus. My examinations confirm, in my opinion, the existence of *Cyclops* [*sensu latu*] with two antennae [i.e. one pair], and thus give the genus *Calanus* the necessary support.

[1845] **Plates 41–43, Crustacés [Copepoda]** *in* **Atlas de Zoologie. Voyages [de la Commission scientifique du Nord] en Scandinavie, en Laponie, au Spitzberg et aux Féröe [pendant les années 1838, 1839, et 1840 sur la corvette** *La Recherche,* **commandée par M. Fabvre]. Publiés par ordre du Gouvernement sous la direction de M. Paul Gaimard. Arthus Bertrand, Éditeur. Firmin Didot frères, Paris.   76 plates.** [From *Atlas Géologique* title page, Cornell University Library.]

Plates 41–43 [Figure legends contain species names only, no text.]

---

[1] Müller's [1785] *Entomostraca,* pl. 19, fig. 8.

[2] [French] *Regne animal,* 2 ed. IV, 157: the one which Mr. Leach designated under the generic name *Calanus,* could indeed form its own subgenus, if it were true that the animal, of which it is the type, did not have second antennae; but has he ascertained this for himself or is he speaking only according to Müller?

Plate 41

Fig. 1. *a–n. Calanus spitsbergensis* Kr. n. sp.
[*Calanus finmarchicus* (Gunnerus)]
[1,*a*. Adult female
1,*b*. Adult female
1,*c*. Rostral area
1,*d*. First antenna
1,*e*. Second antenna (x = base; y = anterior ramus; z = posterior ramus)
1,*f*. Labrum
1,*g*. Mandible
1,*h*. First maxilla (x = "first jaw-pair"; y = "second jaw-pair")
1,*i*. Second maxilla
1,*k*. Maxilliped
1,*l*. Leg 1
1,*m*. Leg
1,*n*. Urosome]

Fig. 2. *a–g. Calanus hyperboreus* Kr.
[2,*a*. Nauplius V (Krøyer's "stage 6")
2,*b*. Nauplius VI (Krøyer's "stage 5")
[2,*b*.¹] First antenna
2,*b*.² Second antenna?
2,*b*.³ Mandible palp?
2,*b*.⁴ First maxilla?
2,*b*.⁵ Caudal segments
2,*c*. Copepodid I (Krøyer's "stage 4"), large form
2,*c*.⁺ Copepodid I, small form
2,*d*. Copepodid II (Krøyer's "stage 3")
2,*e*. Copepodid III (Krøyer's "stage 2")
2,*e*.⁺ Copepodid III, parasitized
2,*f*. Copepodid IV (Krøyer's "stage 1"), parasitized
2,*g*. Copepodid V (Krøyer's "adult") (x = parasite)]

Fig. 3. [*a–c.*] *Calanus quinqueannulatus* Kr. n. sp. [*C. finmarchicus* (Gunnerus)]
[3,*a*. Adult male
3,*b*. First antenna
3,*b*.⁺ Maxilliped
3,*c*. Urosome]

Fig. 4. *Calanus minutus* Kr. n. sp. [*Pseudocalanus minutus* (Krøyer)]
[4. Copepodid V, male]

Fig. 5. *Calanus affinis* Kr. n. sp. [*C. finmarchicus* (Gunnerus)]
[5. Copepodid V]

Fig. 6. *a–k. Calanus cristatus* Kr. n. sp.
[as *Calanus iristatus, lapsus calami*]
[6,*a*. Copepodid V

6,*b*. Anterior margin, dorsal
6,*c*. First antenna
6,*d*. Mandible
6,*d*.⁺ Mandible blade
6,*e*. Second antenna
6,*f*. First maxilla
6,*g*. Second maxilla
6,*h*. Maxilliped
6,*i*. Leg 1
6,*k*. Urosome]

Plate 42

Fig. 1. *a–x. Pontia pattersonii* Tmplt.
[*Anomalocera patersonii* (Templeton), as *Pontia patteronii, lapsus calami*]
[1,*a*. Male
1,*b*. Male
1,*c*. Female
1,*d*. Rostrum
1,*e*. Right first antenna, male
1,*e*.⁺ Right first antenna, male, geniculation
1,*f*. Left first antenna, male
1,*g*. First antenna, female
1,*h*. Second antenna
1,*h*.⁺ Second antenna, exopod
⁺ Labrum
1,*l*. Mandible
1,*m*. First maxilla
1,*n*. Labium
1,*o*. Second maxilla
1,*p*. Maxilliped
1,*q*. Leg 1
1,*r*. Leg
1,*r*.⁺ Leg, terminal spine
1,*s*. Right leg 5, male
1,*t*. Left leg 5, male
1,*u*. Left leg 5, female
1,*v*. Urosome, male
1,*x*. Urosome, female]
[an extraneous 1,*e*.⁺ and 1,*s*., in the center of plate 42, do not seem to relate to these organisms]

Fig. 2. *Calanus caudatus* Kr. nov. sp. [*Metridia longa* (Lubbock) female CV ?]

Fig. 3. *a–e. Ifionyx typicus* Kr. nov. gen. et sp. [*Candacia pachydactyla* (Dana)]
[3,*a*. Mandible blade
3,*a*.⁺ Mandible palp
3,*b*. First maxilla (x = ?)
3,*c*. x = second maxilla; y = maxilliped
3,*c*.⁺ Maxilliped
3,*c*.⁺⁺ Maxilliped, terminal segments
3,*d*. Leg 1
3,*e*. Urosome, female (x = spiniform "limbs")]

Fig. 4. *a–e.* *Thaumatoessa typica* Kr. nov. gen. et sp. [*Thaumaleus typicus* Krøyer]
[4,*a.* Habitus, lateral
4,*b.* Habitus, dorsal
4,*c.* First antenna
4,*d.* Leg 1
4,*d.*⁺ Seta, magnified?
4,*e.* Last thoracic segments and urosome ($x =$ leg 5)]

Plate 43 [Though not strictly of pelagic copepods, this plate is added here for the sake of completeness]

Fig. 1. *a–p.* *Harpacticus uniremis* Kr. n. sp.
[1,*a.* Habitus, lateral
1,*b.* Habitus, dorsal
1,*c.* Rostral area
1,*d.* First antenna
1,*e.* Second antenna
1,*f.* Labrum?
1,*g.* Mandible
1,*h.* First maxilla
1,*i.* Labium?
1,*k.* Second maxilla
1,*l.* Maxilliped
1,*m.* Leg 1
1,*n.* Leg
1,*o.* Leg 5
1,*p.* Urosome]

Fig. 2. *a–p.* *Harpacticus gibbus* Kr. n. sp. [*Thalestris gibba* (Krøyer)]
[2,*a.* Habitus, lateral
2,*b.* Habitus, dorsal
2,*c.* Rostral area
2,*d.* First antenna
2,*e.* Second antenna
2,*f.* Labrum?
2,*g.* Mandible
2,*h.* First maxilla
2,*i.* Labium?
2,*k.* Second maxilla
2,*l.* Maxilliped
2,*m.* Leg 1
2,*n.* Leg
2,*o.* Leg 5
2,[*p.*] Urosome]

Fig. 3. *a–n.* *Harpacticus cronii* Kr. n. sp. [*Parathalestris croni* (Krøyer)]
[3,*a.* Habitus, lateral
3,*b.* Habitus, dorsal
3,*c.* Rostral area
3,*d.* First antenna
3,*e.* Second antenna
3,*f.* Mandible
3,*g.* First maxilla
3,*h.* Second maxilla

3,*i.* Maxilliped
3,*k.* Leg 1
3,*k.*⁺ Leg 1, endopod, terminal segment
3,*l.* Leg
3,*m.* Leg 5
3,*n.* Urosome]

(1848) "Contribution to Carcinology." *Naturhistorisk Tidsskrift,* **series 2, 2, 5: pp. 527–560.**

I. Genus *Calanus* Leach

It cannot be anything but a surprise that such an animal form is still almost completely unknown, which seems to occur through nearly all the seas on the globe, at least in the temperate and cold zones both on this side and on the other side of the Equator, which appears in both the open sea and close to the coast, and which is found in such amazing numbers, that in spite of its small size it forms a not insignificant part of the food of whales. That is the case with animals of the genus *Calanus.* At Spitsbergen I have seen them both in the sea several miles from the coast and in the fjords, in unbelievable numbers; nearly all the birds had their stomachs packed full of them, and it would have been easy to fill whole barrels with them, in spite of their small [individual] mass. I found them only very seldom on Finmark's coast. When I went south, I lost sight of them for a long time, and therefore I almost believed that this genus was only a polar form. But, I was too quick to come to that conclusion, as on the morning of 29 April, 1839, in Engesund (between Bergen and Stavanger) almost at the time of high tide I went to the beach to get a glass of clean water, and with closer inspection found this to be inhabited with a great number of individuals of the genus *Calanus.* Now after a few years, on 6 August, 1844, the same fact was repeated at Hirsholmen in the northern Kattegat, when I got a bowl of seawater one morning at high tide, and after bringing it home found in the water Cirripedia offspring, and countless numbers of Entomostraca, and among those also one little *Calanus* (not an adult animal, but only one just hatched).[1] The form also showed up in the most northerly point in the Kattegat. Toward the west it not only reaches Greenland, but lives also in the sea around Kamchatka.[2] That it also was found in the southern

---

[1] In order not to be misunderstood, I must say, that countless numbers of times I went and got seawater for the animals, which I kept alive to observe, and had never found a single *Calanus* in that.

[2] The species found at Kamchatka is described below. It was brought home by the physician Mr. Schneider from a trip with a whaler, and can be found, just like the other species described here, stored in the Royal Natural History Museum.

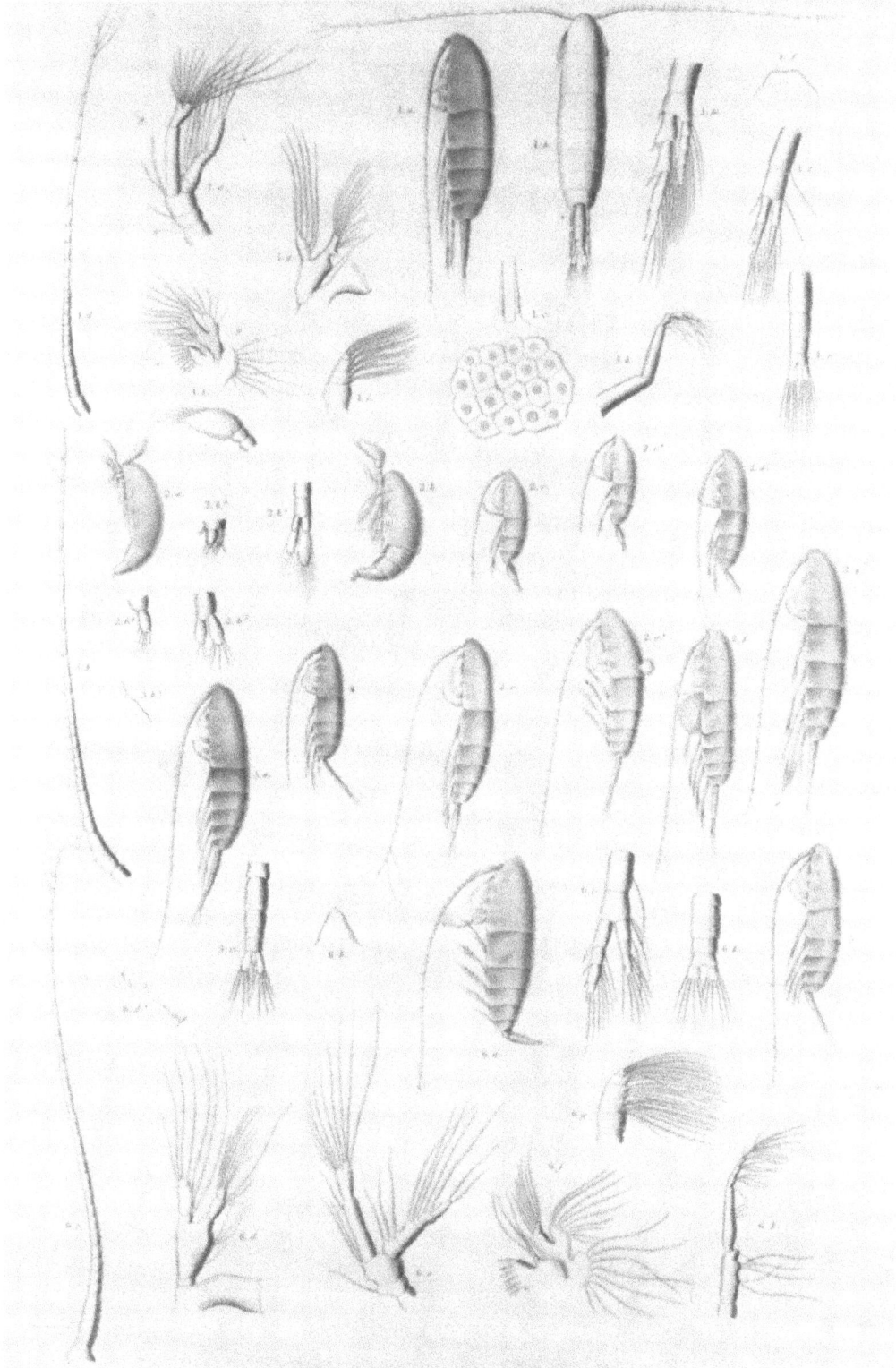

Plate 41 from "Crustacés" in: *Atlas de Zoologie.*

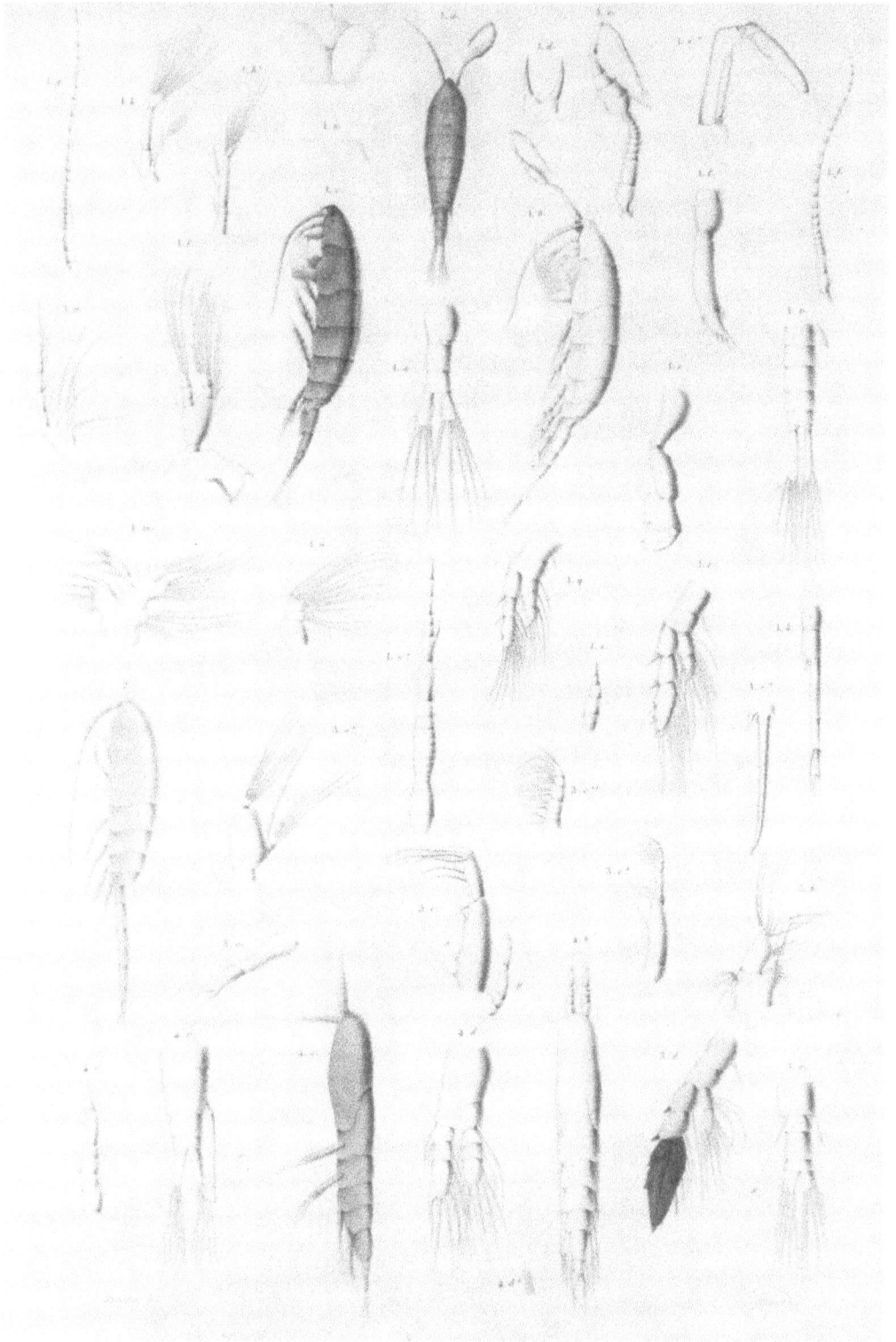

Plate 42 from "Crustacés" in: *Atlas de Zoologie.*

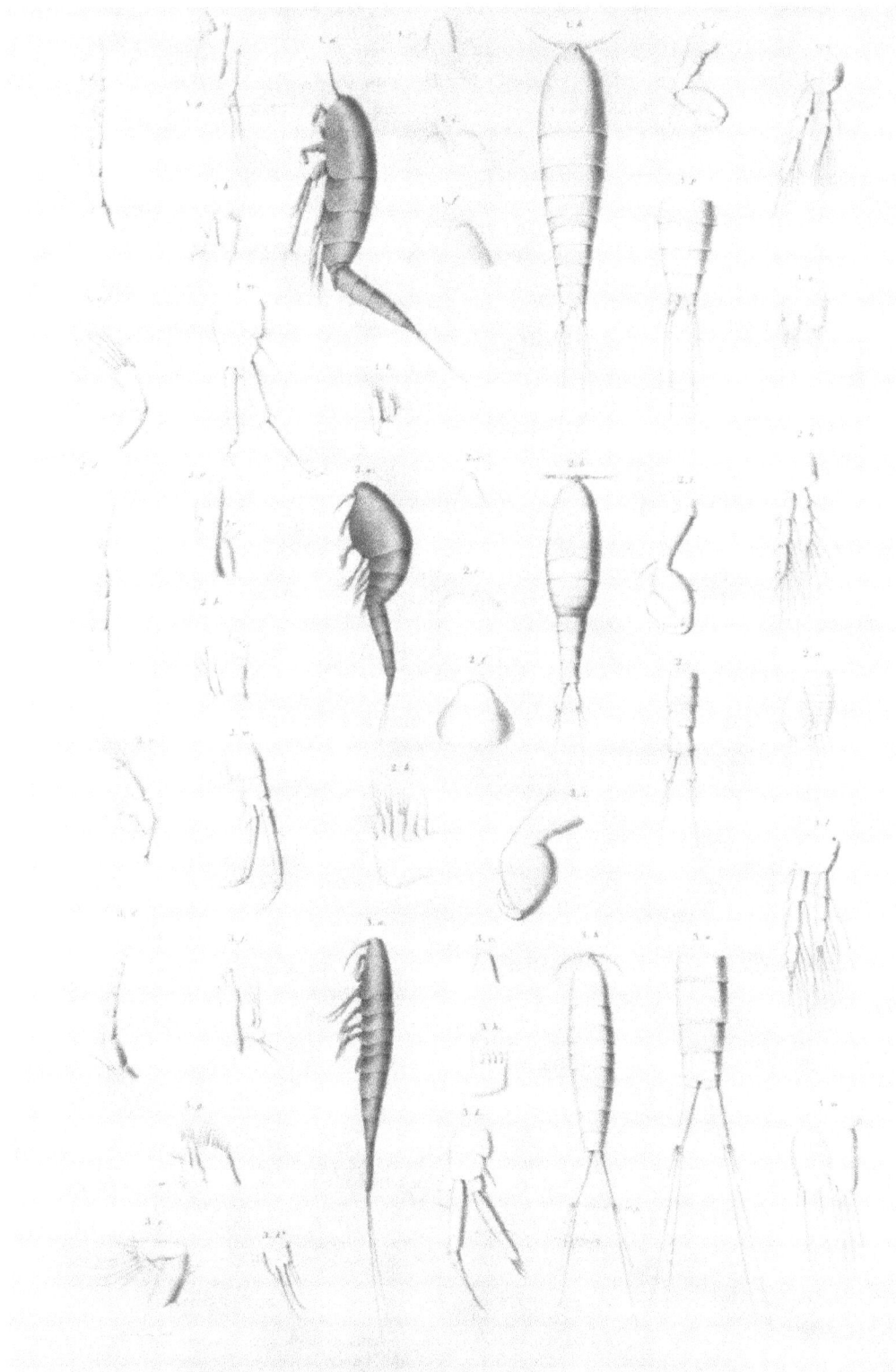

Plate 43 from "Crustacés" in: *Atlas de Zoologie.*

part of the Pacific Ocean and the Atlantic Ocean (42° South), is proved by Roussel de Vauzeme [(1834) *Ann. Sci. Naturelles* (Ser. 2, Zool.) 1: pp. 333–338, pl. 9B]; because the genus *Cetochilus* which was established by him obviously falls in with *Calanus*. And I almost believe that Templeton's [(1836) *Trans. Entomol. Soc. London* 1, 3: pp. 185–198, pl. 20–22] *Calanus arietis*, which was caught in the middle of the Atlantic Ocean, also belongs to the present genus (based on the character of the first antennae), and not to the genus *Cyclopsina* as Milne Edwards indicates (*Hist. d. Crust.* 3: p. 429). I have found a species of genus *Calanus* in the Atlantic Ocean at 30° South close to the coast of Brazil and at 42° North out from Finisterre [Spain]; and I have obtained a second from the East Indian Ocean off the coast of Pinang Island [northern Strait of Malacca].

The first one, as far as I know, who described an animal of this genus, was Bishop Gunnerus of Trondheim, who wrote a short description together with figures of this species under the name *Monoculus finmarchicus* in the *Kjøbenhavn Selskabs Skrifter* (10: pp. 175–176) almost eighty years ago [1770]. Though the figures are not of such nature that the species can be determined from them, they are, however, sufficient not to leave any doubts about the identity of the genus with that of my animals, dealt with a few years ago under the name *Calanus hyperboreus*, and thus further strengthen the description. Müller [1776] took up this species in his *Prodromus* (p. 201 n. 2415) as *Cyclops finmarchicus*; and later in his *Entomostraca* [1785] (pp. 115–116) took it to be identical to a crustacean caught in Drøbakfjord, and therefore placed it together with his *Cyclops longicornis*. Here we must note: (1) that M., to judge from his figures, did not have a full-grown animal in front of him, but a juvenile in one of its last developmental stages; (2) that his description about it was just a few words; and (3) that his figures were very poor and give much less information than those of Gunnerus. I assume in the meantime it is to be unquestionable and obvious, that M.'s species is different from G.'s, but believe that both could pertain to the same genus, *Calanus*, the way I look at them.[3]

Yet a few words must be added about the justification of this genus which was not recognized by Milne Edwards. It was established by Leach [(1819) *Dictionnaire des Sciences Naturelles* 14: pp. 539–540] on the above mentioned *Monoculus* or *Cyclops finmarchicus* without dissection, and its characteristics correspond with Gunnerus's and Müller's descriptions, with the strong development of the first antennae and with the lack of second [antennae]. Latreille [1831, English translation by H. M'Murtrie: *The Animal Kingdom . . . The Crustacea, Arachnides and Insecta.* 3: p. 121. (New York, Carvill)] agrees that (*Regne animal.* 4: p. 157) the genus is valid, if these characteristics are correct, which he doubts with reference to the last and essential part. And his doubts are entirely well founded, as two pairs of antennae actually are present, though the back ones are comparatively small and also are placed rather far back, and were explained differently by the earlier investigators.[4] But there are other reasons for separating the different genera, and among them are those which Latreille himself referred to for *Cyclops castor*, and in following which Milne Edwards has combined this with several species under the genus name *Cyclopsina*. But it [*Calanus*] does not come under that genus either, but belongs instead to Pontiernes family after Edwards's obsolete system, and among these to the genus *Cetochilus* Rouss. de Vauz., which Edwards I believe would have commented on, if he had seen Gunnerus's illustrations. Moreover, the genera *Cetochilus* and *Calanus* also fall together, and, at least according to my impression, the genus name *Calanus* which is much older, even though weakly defined until now, has priority.

1. *Calanus spitsbergensis* Kr. [1845]
[*Calanus finmarchicus* (Gunnerus, 1770), adult female]
[1845, pl. 41 (fig. 1)]

For lack of any better marking in the outer habitus, I shall call the present form the Spitsbergenske; not because it is the only one found in the sea off Spitsbergen (on the contrary, I have gathered five species or subspecies there); nor because it exclusively belongs there, for I have also noticed them from Greenland

---

[3] Whether Müller's species is a *Calanus* or a *Cyclopsina*, or perhaps belongs to a closely related genus, depends on whether it has distinct eyes, and, if so, are there several or only one, as Müller reported. As one generally gives M.'s information much confidence, in some way I consider it my obligation to Science to point out what was reported to me by men, who, I assume, would know the truth about this point. "When M. came back from his excursions to Norway, the well-fed rich man would seek recreation in sleep, while his brother, the poor copper-engraver, spent the evening, and part of the night also, drawing the gathered collection, and on the edge of the drawings were written one or another zoological remark, etc. Often the descriptions were later based only on these drawings and random observations." I

hope, one will admit, that this notice is not irrelevant to Science, and that it would be a wish (though one can hardly hope) that anyone from that time would be alive and able to express himself, be it confirming or rejecting.

[4] In my earlier description of one of the most northern species of the genus ("Grønlands Amfipoder," p. 82 fig.) I also, in spite of much difficulty, had not been able to unravel the truth about that point, but I can mention, as my apology, that my investigation at that time was narrowed down to only a single individual, which we did not dare to cut up. One who is familiar with such undertakings will therefore, I hope, find the imperfections in the description forgivable in some way.

and Iceland. But in shipments from the two last mentioned places it was found only individually, whereas it was by far the most numerous form at the first place.

Among a great many full-grown individuals I have not seen a single one that was over 4.2 mm; by far most of them did not reach that, measured from the edge of the forehead to the tip of the urosome (but excluding the caudal setae).

The color is generally a beautiful rose-red, at times pale, other times deeper.

Just as in the genera *Cyclops, Pontia*, etc. it is also in the present genus that the head and thoracic segments are so closely united, that they nearly amount to just one piece, and that one can therefore say about the animal that it consists only of the body (or cephalothorax) and tail (or urosome). The shape of the cephalothorax represents an elongate ellipse, whose height or width is contained 3 to $3\frac{1}{2}$ times in its length. A nearly imperceptible cross-stripe marks the border between the head and the first thoracic segment; mutually the thoracic segments are somewhat more distinctly separated from each other. The length proportion of the head and the five thoracic segments is almost expressed with the numbers $24 + 12 + 6 + 7 + 5 + 7$. Therefore, the head amounts to almost 2/5 the cephalothorax's length, and that is almost the length of the first three thoracic segments together. The urosome is more distinctly offset from the cephalothorax and much thinner, though it is thicker at the base in the present species than in the other species known to me. The integument is comparatively tough, like leather. The head seems to consist of just one piece; or apparently as in the genus *Pontia*, where it is admittedly two segments, however, in every case it may be agreed that the border between these is very indistinct, if not to say entirely indetermined. In front the head is quite rounded-off in all directions, and sends out downward and slightly backward, trailing side by side, two small rostril filaments, which do not show very distinctly, until the first antennae have been removed. Lying in the same plane, these two filaments, which are separated at the base by a crescent-shaped notch, and which do not reach beyond the first segment of the shaft of the first antennae, run for the most part parallel with each other, but converge slightly near the ends. They are thin, setiform, pointed, and seem to be two-segmented, in that almost the last fifth of each is apparently set off from the upper rostrum by an articulation; I have sometimes found the point of this last segment with a bifurcation. The eyes' relation is much more odd and strange, if the eyes can be found at all. What I mean to admit with certainty, is that there are no differences in the pigment of the body's main color. I will not talk about the individuals that were preserved in alcohol, but instead just refer to the innumerable individuals I have had opportunity to examine alive, in which I have never been able to discover any sign of pigment; just as Gunnerus already drew attention to the lack of eyes in his *Monoculus finmarchicus*. With specimens in alcohol under the microscope one frequently notices one very small, fairly circular, light spot in front of each first antenna's base, which from appearance one could be inclined to assume these spots to be a pair of simple eyes, where pigment had disappeared; with *Calanus hyperboreus* I have explained them earlier like this.[5] But the anatomical examination does not confirm that belief; the spots seem, as well as I can explore, to come just from an accumulation of oil at those places, and entirely disappear by pressing, and the oil oozes out. On the contrary with this species there is a peculiar mass over a very large portion of the head toward the front and particularly dorsally, almost to its posterior edge; it consists of a large number of well-dispersed, small, circular, flat, lens-shaped bodies, each surrounded by a circle of softer substance. All of that large mass, the meaning of which is yet a guess to me, is seen, when it is pressed, to be composed of many irregular, angular figures, which customarily have a five-sided, seldom six- or four-sided outline, and seem to be pointed or conical; I have found the largest diameter of these bodies to be 0.08–0.13 mm; they each are composed of circular bodies of almost 0.04 mm diameter, and in the middle of those appear the above-mentioned, lens-shaped bodies, which seem to be not only very hard but also very elastic, so that I could never break them between two glass plates. The diameter of these runs almost to 0.02 mm. I have often been able to see still smaller concentric circles inside of these.

The first antennae somewhat exceed in length the animal's total length (almost their last two segments project beyond the tip of the urosome, when they are bent back), have also a comparatively very strong construction, and appear to be slightly curved at the base. Moreover, these implements are very seldom encountered entirely whole and uninjured.[6] In the complete condition they consist of a very short shaft and of a long flagellum, which numbers 23 segments. The shaft amounts to only 1/7 or 1/8 the flagellum's length, and has only two segments, and is not too different from the flagellum, so it might seem doubtful whether the name rightly applies to that; but I still find it confirmed as well with analogy as with developmental stages; the first segment is slightly

---

[5] When Roussel mentions eyes in the south-seas species, but passes by their color with silence, and is indefinite about the lenses, I am inclined to believe that he considered those points to be eyes.

[6] I have much less often found complete antennae in this form than in *C. hyperboreus*, which must indicate a greater weakness of these parts in the present species.

pinched at the base; the second longer, cylindrical. The flagellum's first eight to ten segments are much shorter, less distinctly separated; those following increase gradually until the seventeenth or eighteenth, whereupon the last segment again decreases somewhat, but in such a way that the terminal segment is longer but in addition noticeably thinner than the next to the last one; also the last segment is linear, while those preceding have a club form. The third last and the next to the last segment are (in complete condition) each armed with a long, strong, finely segmented or even cross-striped seta, which often is pointed straight out from the side at a right angle; the last segment ends with a tuft of fairly long and very fine setae, that seem to be plumose. The remaining segments are armed at the distal-anterior edge with some fairly long and robust, nearly ribbon-shaped setae.

The second antennae are set fairly far back of the first, and are much smaller in length (they are contained close to five times in the total length), but of a still stronger construction, and represent, just as in the genera *Pontia* and *Cyclopsina,* a pair of especially powerful, two-branched swimming-appendages. The base is three-segmented (the segments' mutual length proportion is almost $3 + 1\frac{1}{2} + 2$), the first segment however is fairly indistinct and indefinitely limited, the second segment is of much larger width than length, somewhat pinched at the base, cut-off at the end, supplied with a long plumose seta at the anterior end; the third segment is rectangular, with two long setae toward the end of the anterior edge. The rami exceed the base in length (comparing almost as 11 to 7), and the anterior is again slightly longer than the posterior one (the comparison about like 11 to 10). It consists of three segments, despite the last of which not only with respect to its small size must be called rudimentary, but also is so indistinctly offset from the second, that only a very strong magnification or a favorable lighting permits recognizing it. The length proportion of these three segments can be expressed by the numbers $8 + 2\frac{1}{2} + 1$; the first is of elongated, linear form, with a pair of fairly long plumose setae toward the end of the anterior edge; the second segment at the end is slightly wider than the first, with the base fairly strongly pinched, of a somewhat rectangular form, yet in this way, that the lowest most posterior angle is cut out so as to take up the little third segment; the free part of the segment's terminal edge is armed with six or seven very long plumose setae, which increase in length from front to back. Also the somewhat pointed third segment has the posterior edge armed with very large plumose setae, five to seven in number. The most posterior ramus is eight-segmented; the segment's mutual length proportion is almost $1\frac{1}{2} + 2 + \frac{1}{2} + \frac{1}{2} + \frac{1}{2} + \frac{1}{2} + 4 + \frac{1}{4}$; the segmentation is also fairly indistinct; the four

middle segments are of much greater width than length; the last segment is entirely rudimentary, scarcely taking in half of the next to the last segment's distal width; all segments are supplied with very long plumose setae on the anterior edge and on the end.

The mouthparts' characteristics are fairly difficult to understand with certainty, because of the softness and the somewhat indeterminate outlines, arising from a weak development of both their epithelium and muscle-system.

The labrum has the form of a wide but short, crescent-shaped or bent-out plate, whose free edge along the middle is armed with a few short setae.

The mandibles have a comparatively important size and a robust form, and consist of the mandible proper and the palp. The first is elongated, somewhat enlarged toward the base, strongly laced together at the middle, and again toward the free or inner end enlarged almost into a fan- or ear-form. The chewing surface is represented in the middle of a somewhat sunken oval, whose edges are armed with teeth in this way, that the front teeth are larger and widely separated, the back [teeth] smaller, closer together, and nearly seta-like. The number of teeth is difficult to determine; I have observed almost twenty pieces. The palp is almost the same size as the mandible proper, and consists of a two-segmented basal part, from which extend two small, two-segmented branches or rami (implements for water movement). The basal part's first segment is not only much shorter in comparison with the second, but also very thin, though of larger width than length, and looks only like a constriction between the mandible and palp. The second segment, which is four or five times as long as the first, has an irregular heart shape in that it is somewhat pinched below, widens out toward the end, and also has a fairly deep cutout between the two branches; a pair of long setae extends from the inner anterior side. The two segments of the inner branch or ramus have almost the same length, or the first one is only insignificantly longer than the second, but separates itself from it with a significant prolongation, arising from the inner side, which is bluntly rounded off and thick at the end, and from whose base three long plumose setae extend forward. The second segment is robust, blunt, rectangular, somewhat wider at the end than at the base, and on the directly cutoff distal edge is supplied with eight very long plumose setae (longer than the entire mandible with palp). The outer ramus is almost the same length as the inner and just like that shorter than the basal part's second segment. Besides the two distinct, robust, blunt, rectangular, mutually almost the same size segments, whereof it consists, there still seems to be present a rudimentary distal segment. This branch's inner edge, right from the base to the point, is occupied by six plumose setae that are still longer than those situated on the inner ramus.

The first maxillae are formed of five small plates [7] which are so close in alliance with each other that because of their small size it becomes difficult to separate them anatomically. The two innermost and first of these plates I regard as constituting the first jaw-pair: the inner, larger, and wider is the actual jaw, the smaller is the outer palp; the first along the inner edge and at the end appears armed with about ten very large and strong spines, which seem to be set in two rows and along the posterior edge are supplied with several very short setae, also apparently plumose, or intermediate between plumose- and saw-form. The palps have a pair of long plumose setae at the end and five or six along the inner edge. The remaining three plates must be regarded as constituting a second pair of jaws: the innermost is the actual jaw, the second the palp; the third represents the flagellum, or a kind of a gill-plate. The jaw consists of three, yet fairly indistinct, robust, short and wide segments, that decrease slightly in length from the first to the last; from the last segment's flatly cutoff distal edge extends about six very long, inwardly turned plumose setae, and one from each of the two preceding segments' inner edge. The palp shows no segmentation, it is slightly pointed toward the end, and bears five or six very long plumose setae, nearly pointing straight out at a right angle on the outer edge. The crescent-shaped flagellum is far back and outside the other two plates, and has the edge supplied with eight long plumose setae, pointing out toward the back.

The second maxillae are small, of fairly robust form, conical, somewhat forward-curved, abundantly supplied with extremely long plumose setae (they considerably exceed the maxillae in length), their characteristics are like a swimming-appendage, or like an appendage to set the surrounding elements in motion. They consist of five segments, whose mutual length proportion can almost be expressed by the numbers: $6 + 4 + 4 + 1 + 1$. The first segment shows, looking from the side, a fairly regular rectangular outline of almost the same length and width; the second segment also is rectangular, but of greater width than length; on the anterior side extend two fairly large knots or cylindrical, terminally rounded prolongations, each of which is armed with two very long plumose setae; the third segment is almost of the nature of the second; the fourth segment has a greater width than length, and shows toward the front just one, but in proportion a very large knot, supplied with three long plumose setae; the small fifth segment is supplied with

a fairly rudimentary knot; maybe there is yet a sixth segment present; however, if that is the case it is completely rudimentary and indistinct. Altogether the number of the long plumose setae amounts to about twenty; because of their length and close position they appear just like a brush; at the base they are so thick that they look almost like forklike ramifications of the knots from which they arise.

The maxillipeds are more than twice as long as the second maxillae, and constitute almost 1/5 or 1/6 the total length; in contrast they are of thinner form, though strong; they are curved forward and make a swimming-appendage, though less complete than the second maxillae. One can distinguish here between a basal part and a ramus; the basal part consists of two or maybe three segments, but the first segment in this case is very small and indistinct. The basal parts' two distinct segments are of elongate form, mutually almost the same length, but the first is much thicker, at the base on the anterior side supplied with a knot or swelling, the distal posterior edge obliquely cut off, while the second segment has a more regular cylindrical form; both are supplied with several long plumose setae on the anterior side. The ramus does not have half the length of the basal part, and consists of five segments, of which the last however is nearly rudimentary; the segments' mutual length proportion approximates $4 + 6 + 3 + 3 + 1$. The form of the ramus is fairly robust, gradually tapering to a point; its forward edge is supplied with plumose setae.

The thoracic segments, which always number five in adult animals, and which each carry a pair of strong swimming-feet, give only cause for a negative remark; namely, that on the posterior outer edge of the fifth segment there is not such a thornlike prolongation, which always seems to be found on animals of the genus *Pontia*.

The length of the first pair of feet constitutes somewhat less than 1/5 but more than 1/6 the total length, or is almost as long as the maxillipeds. The length of the two-segmented basal part is of the same proportion to the length of the outer ramus as nine to eleven, and the outer ramus's length is again in proportion to the inner ramus as eleven to eight. The basal parts are therefore slightly shorter than the outer and slightly longer than the inner rami; its first segment is twice as long as the second. Each ramus consists of three distinct segments; the mutual length proportion of the segments of the outer ramus is almost $3 + 3 + 5$. The first segment of the inner ramus is comparatively large (longer than the two following segments together), the others very small. Both rami are abundantly supplied with long plumose setae on the inner side (I do not dare to give the exact number); also the two segments of the basal part each have a long plumose seta on the inner side toward the end; the outer ramus is armed with spines

---

[7] Milne Edwards, who has observed only three of these plates in his genus *Pontia*, refers to all of them as first jaw-pair, and assumes therefore that the first pair of jaw-feet [second maxillae] is a second pair of jaws.

at the end of the segments on the outer side (one on each of the first two segments, three on the last segment); these spines are very large and strong on the first two segments.

The second pair of feet is slightly longer than the first, and shows some deviation in the parts' mutual relation; the length of the basal parts is related to the outer ramus as 2 to 3½, and the outer ramus to the inner as 16 to 9. The first segment of the basal part is almost three times as long as the second. The length proportion of the outer ramus's segments is 4 + 4 + 9. Of the spines, which are set on the outer side of these segments, the last one on the end is especially large and strong; on the inner side this ramus has seven long plumose setae (one on the first, one on the second, and five on the third segment). The inner ramus's first segment is very small, not half as long as the second; the third segment is longer than both preceding ones together, with a ledge, from which plumose setae extend, as if separated in four or five smaller segments; the proper explanation would, no doubt, be just to regard that part as a single segment. The number of plumose setae on the inner ramus is eleven: one on the inner side of the first segment, two on the second segment, eight on the third segment, of which six are on the inner side and on the end, the other two on the outer side.

The third pair of feet is slightly longer than the second but almost of the same form and proportion. The only deviation, I have observed, is that the outer ramus's third segment increases somewhat in length (the segments' mutual proportion is therefore almost: 4 + 4 + 10). The number of spines and setae is as in the preceding pair of feet.

The fourth and fifth pairs of feet correspond in form with the third, except the fourth pair is slightly longer than this, whereas the fifth pair is shorter than the three preceding and only very slightly longer than the first.

One can also remark about the feet, that in resting position they all are customarily directed straight backward. In that state the first pair reaches with the point of the setae not quite to the end of the next to the last thoracic segment; the second pair almost to the end of the first urosomal segment; the third pair not quite to the end of the fourth urosomal segment; the fourth pair beyond the end of the caudal rami (but not to the end of the caudal setae), the fifth pair just almost to the middle of the caudal rami.

The urosome or tail, whose length constitutes almost ¼ the total length, consists in the adult animal of four segments and two caudal rami. The length proportion of the urosome's parts can fairly accurately be expressed like so in numbers: 8 + 5 + 3 + 3 + 5. The first four numbers designate the urosomal segments, the fifth the caudal rami. This proportion seems to be constant in this species; at least I have, by measur-

ing a series of individuals, found it unchanged and the same in all, or yet as close as possible. The width of the urosomal segments decreases gradually, but just very slightly; the first segment has, as mentioned, almost the same length as the two following taken together; it is of a somewhat thick and robust form, fairly strongly swollen on the ventral surface, and appears, where it is cut off from the thoracic part, thicker at the base than at the end, whether one looks at the animal from the side or from above; and as that relation makes it much easier to understand the differences between this and several similar species, I have included that in the diagnosis, but must therefore now, so not to be misleading, remark, that the actual urosomal segment, as in the other species, appears pinched at the base, where it separates from the last thoracic segment, but only for a short distance, which is the reason that it had not been noticed earlier; the second segment is longer than the third, which almost has the same length as the fourth segment; the caudal rami are somewhat longer than the fourth segment, are somewhat linear (or narrow and of equal width), with six partly very long setae at the end, whose length proportion however I cannot determine for sure, as I have always found them in a more or less injured condition. The innermost [seta] on each ramus is, however, the shortest and those in the middle the longest. At the end of the urosome's first segment on the ventral surface toward its outer angles are observed two circular points, noticeable with a darker color than the rest of the body; it is reasonable to designate these the openings of the reproductive organs. By pressing, one can notice in the same segment toward the base or toward its anterior edge two small, bladder-shaped, oval organs. The fourth segment has only a very insignificant notch in the middle of the posterior edge, where the anus is located.

2. *Calanus hyperboreus* Kr. [1838]
Krøyer: "Grønlands Amfipoder," p. 84 onward, and pl. 4, fig. 23
[1845, pl. 41 (fig. 2)]

I have found this species at Spitsbergen and several places on the Norwegian coast; moreover I have recently examined a few individuals sent down from Greenland and Iceland. It is also reasonable to believe that this form happens to be found in the Kattegat, which I cannot decide for sure, as I have only seen developmental stages of this last.

It reaches a length of up to 8.5 mm, but that size seems to be unusual. On the other hand it is as a rule somewhat larger than the preceding species.

The color is at times entirely white, at times white with the first antennae rose-red, at times also entirely red. I am inclined to believe that the first color is the usual.

In general the form seems to be slightly thicker

than in the preceding species, so that the height of the cephalothorax amounts to almost 1/3 the length, while in *C. spitsbergensis* the cephalothorax is almost four times as long as it is high. Yet the mentioned thickness must not be taken as constant in the present species; one quite often finds individuals, which are just as thin as the preceding species.

On the whole the likeness is complete in nearly all the parts, and the few diversities not very constant, so it is very difficult to explain sufficient distinctions between *C. hyperboreus* and *C. spitsbergensis*, and thus I establish them as two species only with much doubt, and leave to the future to deny or confirm my assumption. I cannot actually find any constant distinctions, other than the length proportion of the urosomal segments and caudal rami, and the form of these parts, particularly the first segment. The length proportion is nearly expressed by the numbers $3 + 7 + 4 + 5 + 4\frac{1}{2}$. So it is the first segment that is the shortest, the second the longest. As far as the form is concerned, the first segment is very distinctly and conspicuously pinched together at the base.

3. *Calanus minutus* Kr. [1845]
[*Pseudocalanus minutus* (Krøyer, 1845)]
[1845, pl. 41 (fig. 4)]

I have found only five or six specimens of this form among *C. spitsbergensis* and *C. hyperboreus* from the sea on Spitsbergen's west coast.

It is distinguished from the mentioned species by its much smaller size, because its length does not seem to go over 1.6 mm, and more often perhaps it does not come near that measurement.

The color is reddish, almost like that in *C. spitsbergensis*.

In form this species looks exceptionally much like the two first mentioned and nearest to *C. hyperboreus*. It differs in the nature of the first antennae, which in length are very distinctly less than the animal's total length (almost by 1/5). Bent backward they reach almost to the end of the third urosomal segment.

The length proportion between the head and the five thoracic segments can almost be expressed by the numbers: $9 + 4 + 2\frac{1}{2} + 2\frac{1}{2} + 2 + 1$. Such, which here is characteristic, is the insignificant development of the fifth thoracic segment, which can almost be called rudimentary, and, particularly on top, it is entirely hidden by the preceding segment; whereas it shows from the sides and from the ventral side.

The feet distinguish themselves by a somewhat unusual shortness: the first pair reaches almost to the end of the second thoracic segment, the second pair to the end of the fifth thoracic segment, the third pair to the end of the second urosomal segment, the fourth pair almost to the middle of its fourth segment; but the fifth pair, whose shortness provides a good specific characteristic, reaches only to the end of the second urosomal segment.

The urosome is somewhat longer than in the closely related species, in that it constitutes nearly 1/3 the total length, and almost has the same length as the head. The proportion between its four segments and the caudal rami is almost: $2 + 5 + 4 + 5 + 4$. The first segment appears strongly pinched together at the base, when it is looked at from the side, just as in *C. hyperboreus*. Finally the tail itself is distinguished, in that it is most frequently bent strongly backward.

At the first examination I had been inclined to regard this form as a developmental stage of *C. hyperboreus*, which it agrees with in many respects; but the remarks which I will make below on the development in the genus *Calanus* seem to substantiate its qualification to be considered as a particular species.

4. *Calanus affinis* Kr. [1845]
[*Calanus finmarchicus* (Gunnerus, 1770), juvenile]
[1845, pl. 41 (fig. 5)]

Up until now I have seen only three specimens of this form, all caught in the sea at Spitsbergen.

It stands so close to the two first mentioned species, that it is only with doubt that I separate it from these.

Its size runs only to about 2.1 mm or a little more.

From specimens not preserved in alcohol, the color seems to be the usual beautiful light red.

As it is only the urosome's characteristics, besides its small size, that separates it from *C. spitsbergensis* and *C. hyperboreus*, I limit myself, to avoid unnecessary details, only to bring out these characteristics. It consists, as usual, of four segments and two caudal rami, but the length proportion of these parts is almost $3 + 3 + 8 + 10 + 7$; both the first two segments are also much shorter, and it is the fourth that here is the longest. It must be left to future investigators of the live animals, either to find more distinguishing characteristics for this species, or perhaps to show its identity with one of the first species.

5. *Calanus quinqueannulatus* Kr. [1845]
[*Calanus finmarchicus* (Gunnerus, 1770), adult male]
[1845, pl. 41 (fig. 3)]

I have found a few specimens of this form among *Cal. hyperboreus* from Greenland.

The size can be estimated to be almost, or close to, 4.2 mm.

The color of those specimens preserved in alcohol is fairly dark red or reddish brown; from that I do not believe that I could draw any certain conclusions about the color of the living animals.

What distinguishes this species is the length and characteristics of the antennae, and the number of segments on the tail and its nearly always forward curved direction. Yet observing with closer examination there are several other oddities.

Among these represented first is the length of the head, which is almost the size or even a little larger than the length of the five thoracic segments together, and constitutes almost 2/5 the animal's total length. The mutual length proportion of head and thoracic segments can be expressed by the numbers: $14 + 4 + 3 + 2\frac{1}{2} + 2 + 2\frac{1}{2}$. The head is very strongly bulged outward in front and the rostrum directed backward in a fairly high degree.

The first antennae have a more prominent length than in some of the other species known to me, and are considerably longer than the animal. In their general backward-directed position, they project by almost the last five segments beyond the tip of the tail. Also they are somewhat thinner than usual, and remain distinctive with their peculiar armament, in that the segments not only are supplied with many large and small outwardly directed setae, but also with fairly robust, leaf- or ribbon-shaped, curved organs, that have almost the same length as the segments to which they belong, and for a large part are themselves two-segmented. The shaft is contained not more than six times in the antenna's length; its first segment is not pinched at the base, but expanded, or wider than at the end; the second is of elongate-oval form. The flagellum's first five segments are not as short as in *C. spitsbergensis* and *C. hyperboreus*; it is only the sixth and seventh that here are much shorter; the last segment is not longer than the next to the last.

On the second pair of antennae the setae are much stronger and more definitely developed as plumose setae than in the preceding species.

The first segment of the inner ramus of the mandible palp is not expanded or lengthened on the inner side; the second segment is comparatively very large, and is rounded at the end, armed with six plumose setae. The lengthening of the base, where the outer ramus starts, is very strongly developed; the first segment of the ramus is nearly circular, and is armed with a pair of smaller setae on each side; the second [is armed] at the end with nine large plumose setae.

The urosome, whose length constitutes almost $\frac{1}{4}$ the total length, is distinguished by its fairly strong bent-forward direction (a condition that, if not entirely constant with this species, is, however, by far the most common) and consists of six segments (the tail included) whose mutual length proportion is: $3 + 6 + 4 + 3 + 3 + 5$.

6. *Calanus cristatus* Kr. [1845]
[1845, pl. 41 (fig. 6)]

It is this species, to which I already have had occasion to refer to above, that belongs to the sea off Kamchatka. The few specimens I have seen were all less than well preserved.

This species seems as a rule to have a length of almost 8.5 mm, and can thereby be regarded as the largest of the species known up to now in this genus.

The color of these specimens preserved in alcohol is entirely white.

Their form seems somewhat thicker and more robust than is customary in the species of this genus, although this depends only on the shortness of the urosome, in that the greatest diameter just amounts to about $\frac{1}{4}$ the length of the cephalothorax, or that [the urosome] is contained more than $3\frac{1}{2}$ times in [the cephalothorax]. The mutual proportion of the length of the head and the five thoracic segments can be expressed almost like this: $17 + 8 + 4 + 4 + 3 + 2$.

The head, which is about the same length as the thorax, is distinguished by its slightly pointed form in front and with a fairly strong, protruding longitudinal-crest in the middle of the anterior edge. By this crest it is very easy to separate the present species from the northern species known to me.

The rostrum is the usual forklike cleft; its rami are smaller than 1/20 the total length (almost the same as 1/22), and are slightly divergent at the end; their greatest mutual distance between is almost like half of their length; their form is slightly robust, compared with *C. hyperboreus* etc. I have not been able to discover any sign of segmentation.

The first antennae, which slightly exceed the total length, I have found to be formed of a 2-segmented shaft and a 23-segmented flagellum. The shaft constitutes almost 1/12 the length of the antenna, and is just as long as the flagellum's first three or four segments combined. These two segments are in mutual proportion almost as 1 to 5. The first therefore is very short, obliquely cut off, and of much greater width than length. The second, however, is of an elongate form, more than twice as long as wide, cylindrical, and cut off evenly at the end. The first ten segments of the flagellum are very short and are also very indistinctly separated. The proportion of increase and decrease of the segments is almost like that of the Spitsbergen species. The setae of the last segments were damaged in all of the specimens I studied, so I am unable to describe their normal characteristics.

The length of the second antennae is contained nearly $6\frac{1}{4}$ times in the total length and more than seven times in the length of the first antennae. When the length of the base is represented by the number 5, the anterior ramus can be designated by 9 or $9\frac{1}{2}$ (it is also nearly twice as long as the base); the posterior one by 7. The base consists of three robust and somewhat shapeless segments, of which the third is the widest; the anterior ramus is formed of a long, linear segment, which at the end carries a fairly wide, heart-shaped cut-out leaf (of half the length of the first segment); the posterior tip of the heart is slightly longer than the anterior one, and seems to consist of a separate segment. The posterior ramus consists of seven or eight segments: first a short but wide segment; next a segment which is almost twice as long as the preceding; next four small segments, which together do not noticeably exceed the second; finally a long segment, which even exceeds the second segment somewhat in length, and at the end seems to carry an entirely rudimentary and indistinct segment; (the length of these eight segments can almost be estimated as follows: $2 + 4 + 1 + 1 + 1 + 1 + 6 + \frac{1}{2}$). I have found three swimming-setae extending from the anterior side of the base, to the contrary none from the first segment of the anterior ramus, seven from the point of the anterior end-tip, five from the posterior one; one from the anterior side of the first segment of the posterior ramus, three from the second, one from each of the next four, and finally three from the tip of the last segment.

The mandible shows much likeness to the same part in the Spitsbergen *Calanus*; the base of the mandible proper is very indistinctly separated from the thick fan- or ear-shaped part, on the inner edge of which I have counted over twenty pointed denticles, distributed somewhat irregularly in two rows; close below the uppermost of these is a rounded-off knob. The palp is just as long as the width of the mandible proper. The first segment of the base can be described by the number 1 or $1\frac{1}{2}$, when the second segment is designated by 10, the outer ramus by 7, the inner by $6\frac{1}{4}$. The first segment of the outer branch is much shorter than the second, whereas the proportion in the inner branch shows just the opposite. So that the actual mouthparts do not show any remarkable deviation from those described for *C. spitsbergensis*.

The second maxillae have nothing remarkable in their form; they distinctly show five segments, of which the fourth and particularly the fifth are very small; those tips extending from the anterior edge are here rather difficult to observe. On this maxilla's anterior edge and on the end are more than twenty long plumose setae.

The maxillipeds, whose length amounts to almost 2/5 the total length, are distinguished by the thickness of the first segment of the base in comparison with the second, whereas they have almost the same length; the end of the first segment is very obliquely cut off (in direction backward and inward) and on the inner edge is armed with four or five long plumose setae, that extend from two or three small protruding knobs; on the second segment I have observed only one pair of setae of plain form. The length of the flagellum is almost the same as half of the base or one of its segments, and consists of five segments, all distinct (though the last is fairly small) and each is armed with a pair of plumose setae.

The thoracic feet are of much stronger construction, but are fairly short. The length of the first pair is contained almost six times in the total length; the two-segmented base is almost the same length as the outer flagellum; its first segment is twice as long as the second. The three segments of the outer flagellum compare mutually almost as $3 + 2 + 3$. The inner flagellum compares to the outer nearly as 6 to 7, or reaches almost to the middle of its third segment; the length proportion of the segments is almost $5 + 2 + 5$. The spine- and the seta-armature are almost like those of the other species of the genus.

The length of the urosome is contained almost five times in the total length; it is linear, fairly robust (the thickness amounts to almost 1/3 the length). The mutual length proportion of the segments can almost be described as follows: $3 + 6 + 4 + 3 + 3$, where the last number designates the caudal rami. The first segment is strongly pinched at the base, at least twice as wide as it is long; the second segment is of almost the same length as width, a fairly regular rectangle, yet slightly narrower at the base; the third segment is a rectangle, almost half again wider than it is long; the fourth segment is likewise of much greater width than length, but at the end toward the sides it is slightly obliquely cut off, where the caudal rami are connected.

The caudal rami are contained fully five times or even slightly more in the length of the four urosomal segments; they are of oval form, the length half of the width, widely separated from each other (the space between them is sometimes even greater than the width of each caudal ramus).

[7.]　*Calanus caudatus* Kr. [1845]
[*Metridia longa* (Lubbock) female CV ?]
[1845, pl. 42 (fig. 2)]

Among the developmental steps of *Calanus hyperboreus* were found several individuals of one form, which, although they were distinctly characterized as a developmental step, by having only four pairs of feet, appeared to be quite different from *Calanus hyperboreus* and closely related species, and could not be included with any of these. It is the long thin tail, that at first glance makes this form recognizable, which I have sought to suggest with the species name. The length of the smallest individual is almost 0.5 mm or slightly more, of the largest up to 1.1 mm [3.2 mm in the table on p. 46].

The head is not divided into two segments, but represents only one mass; its length constitutes almost 1/3 the animal's total length. The first antennae are of a very insignificant size compared to the other forms of this genus; [8] they extend only slightly beyond

----

[8] In the beginning I did not pay any attention to this peculiarity, because I believed the antennae were mutilated; but when the antennae of all the individuals, which I examined, were almost the same length, and moreover, when both an-

the last thoracic segment; and their characteristics are as usual.

The thoracic segments are four in number, just like the feet; and they appear much more distinctly separated from each other than in the other calanids. The length of the reach of the four thoracic segments together is somewhat less than the length of the head, and constitutes ¼ the animal's total length.

The tail or urosome is very long (it constitutes 2/5, or even slightly more, the animal's total length) and is comparatively thin. It consists of six segments in addition to the caudal rami; the first segment is somewhat shorter than the second (the longest), but is almost the same length as the third; the fourth shorter than the first and third, the fifth slightly longer than the fourth; the sixth segment the shortest. The caudal rami are almost of the usual proportions.

As I was accidentally deprived of my specimens of this form, I have not even been able to renew my examination of them, and therefore cannot give a more detailed description.

I will now give a diagnosis and a tabulated summary of the aforementioned species, and include that single specimen found off the coast of Brazil, *Calanus carinatus*. They can all be found deposited in the Royal Museum's systematic crustacean collection (except for the lost *C. caudatus*), and I have delivered figures of them to the French Expedition-Bureau (but not of *C. carinatus*).

1. *Calanus spitsbergensis* Kr. [1845]
[*Calanus finmarchicus* (Gunnerus, 1770), adult female]

[Latin] *Head* distinctly shorter than thorax. *First antennae* slightly longer than body. *Second antennae* equal about one-fifth length of animal or first antennae. *Urosome* with four segments, with rather thick bases not at all compressed. Urosome equals about one-fourth of animal's length; length of segments and caudal rami approximately indicated by the following numbers: $8 + 5 + 3 + 3 + 5$. *Length* of adult animal seems rarely to exceed 4.2 mm.

2. *Calanus hyperboreus* Kr. [1838]

Head distinctly shorter than thorax (approximately equaling length of the three anterior thoracic segments together). *First antennae* longer than animal's length (by about the last two or three segments). *Second antennae* equal about one-fifth length of animal and first antennae. *Urosome,* composed of four segments in addition to caudal rami, equal to about one-fourth of animal's length; length of segments and caudal rami approximately indicated by the following

tennae of the same individual showed the same length and finally when no injury could be detected, all doubts about their true state were removed.

numbers: $3 + 7 + 4 + 5 + 4½$. First segment compressed at base. Length of animal reaches 8.5 mm.

3. *Calanus minutus* Kr. [1845]
[*Pseudocalanus minutus* (Krøyer, 1845)]

Head shorter than thorax. *First antennae* distinctly shorter than animal's length (by about one-fifth), not going beyond second urosomal segment when flexed backwards. Fifth thoracic segment generally rudimentary, with top part hidden. *Fifth foot* shorter than usual, scarcely equaling one-seventh of animal's length, nor does it reach beyond second urosomal segment. *Urosome* equals about one-third of animal's length, composed of four segments with two caudal rami; length of these parts approximately indicated by the following numbers: $2 + 5 + 4 + 5 + 4$. First urosomal segment compressed at base. *Length* of adult rarely exceeds 1.6 mm.

4. *Calanus affinis* Kr. [1845]
[*Calanus finmarchicus* (Gunnerus, 1770), juvenile]

Head shorter than thorax, but longer than urosome, equals about one-third of animal's length. *First antennae* exceed length of animal (generally by two last segments, or last segment alone). *Urosome* composed of four segments in addition to caudal rami, equals about one-fourth of animal's length. Length of segments and caudal rami approximately indicated by the following numbers: $3 + 3 + 8 + 10 + 7$. Length of animal approximately 2.1 mm.

5. *Calanus quinqueannulatus* Kr. [1845]
[*Calanus finmarchicus* (Gunnerus, 1770), adult male]

Head as long as or longer than thorax. *First antennae* distinctly longer than animal's length (by about one-sixth, or by 4 to 5 last segments beyond tip of tail), setose and full of small appendages. *Second antennae* equal one-sixth of animal's length. *Urosome* makes up one-fourth of animal's length, five-segmented, *slopes forward.* Length of segments and caudal rami approximately indicated by the following numbers: $2 + 7 + 4 + 3½ + 2½ + 5$. Length of animal rarely reaches 4.2 mm.

6. *Calanus cristatus* Kr. [1845]
[as *Calanus iristatus* Krøyer, 1845]

Head shorter than thorax, armed with longitudinal crest on anterior edge. First antennae barely longer than body. Second antennae not quite one-sixth of animal's length, nor one-seventh of length of first antennae; posterior ramus of these antennae distinctly shorter than anterior. Urosome makes up about one-fifth of animal's length; length of segments and caudal rami approximately indicated by the following numbers: $3 + 6 + 4 + 3 + 3$. First urosomal segment compressed at base. Length of animal reaches 8.5 mm and more.

7. *Calanus caudatus* Kr. [1845]
[*Metridia longa* (Lubbock) female CV ?]

Head longer than thorax. First antennae do not equal animal's length, scarcely surpassing cephalothorax in length. Urosome slender, very long (scarcely less than half length of the animal), composed of six segments in addition to caudal rami; length of these parts approximately indicated by the following numbers: $3 + 5 + 3 + 2\frac{1}{2} + 3 + 2 + 2\frac{1}{2}$.

8. *Calanus carinatus* Kr. [new species]
[*Calanoides carinatus* (Krøyer, 1848)]

Head shorter than thorax, armed with longitudinal crest or keel on anterior edge, which also extends along a large part of the back. First antennae distinctly shorter than body, not reaching penultimate urosomal segment. Urosome equals one-fourth, or slightly more, of animal's length, and consists of four segments in addition to caudal rami; length of these parts approximately indicated by the following numbers: $6 + 4 + 3 + 2 + 3$. Length of animal equals 3.2 mm.

[Danish] On the Development of the Genus *Calanus*.
[1845, pl. 41 (fig. 2)]

After having decided, according to the analogy of closely related genera, which forms could be regarded as peculiar to the mature animal of every species, it followed of itself so to speak, which [forms] could be considered as developmental steps. As I took upon myself the tedious work of gathering several thousand individuals and to scrutinize the species, piece by piece, under the microscope—it was a task I found very difficult, but perhaps not entirely wasted— I gradually was able to distinguish all the larvae. Later I went through them again, to divide them into stages. And in this way I have, alone with the help of specimens in alcohol, attained a fairly extensive and, I must say, reliable knowledge of the transformations which this genus has to undergo. But in accordance with the way the examination was carried out, and I hold it to be correct, in representing it to go backward, if I dare say, or slide gradually from an older to a younger age. This presentation is especially about the species *Calanus hyperboreus*.

1. The larva, which is most like the adult animal, I have found to vary in length from 3.2 mm to almost 5.3 mm. The thorax has five segments and five pairs of feet; the urosome on the contrary has only three segments not counting the caudal rami. The mutual length proportion of these last can almost be described by the numbers $2 + 3 + 5 + 3$. It is conspicuous that it is the third, longest segment, designated here by the number 5, whose partition into two still stands out, and which shows the only difference

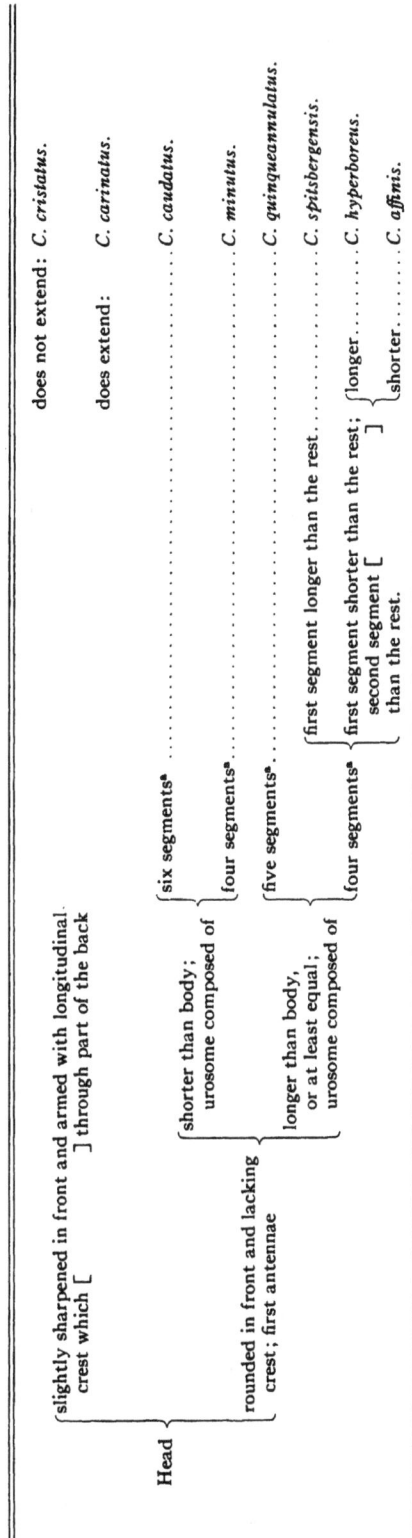

Head {

slightly sharpened in front and armed with longitudinal crest which [ ] through part of the back {

does not extend: *C. cristatus.*

does extend: *C. carinatus.*

rounded in front and lacking crest; first antennae {

shorter than body; urosome composed of {

six segments[*] ......*C. caudatus.*

four segments[*] ......*C. minutus.*

longer than body, or at least equal; urosome composed of {

five segments[*] ......*C. quinqueannulatus.*

four segments[*] {

first segment longer than the rest ......*C. spitsbergensis.*

first segment shorter than the rest; second segment [ ] than the rest. {

longer......*C. hyperboreus.*

shorter......*C. affinis.*

[*] in addition to caudal rami.

between the larva and the adult animal. On the whole, as far as I have observed, the greatest similarity is in the number of the segments of the first antennae.

2. In the young of the nearest preceding developmental step we find five thoracic segments, but only four fully developed pairs of feet, in that the last thoracic segment possesses only one almost unnoticeable little rudimentary foot, about whose true characteristics, however, I have not yet found a clear explanation. The mutual length proportion of the head and thoracic segments can almost be expressed by the numbers $12 + 7 + 4 + 4 + 2 + 3$. The urosome constitutes more than $1/5$ but not $\frac{1}{4}$ the total length. It consists of only two segments not counting the caudal rami. Their mutual length proportion can be estimated to be $2 + 6 + 2\frac{1}{3}$. The antennae are comparatively strong, nearly robust; they exceed the total length only by the last two and almost half of the third last segment, but that happens, because of this segment's strong development, almost $1/6$ the total length. I have found only twenty segments, not counting the shaft. The length of the animal in this stage amounts to almost 1.5 to 2.1 mm.

3. In the third stage in reverse order the larva shows four thoracic segments, three pairs of thoracic feet (the fourth segment lacks feet), and a urosome consisting of two segments not counting the caudal rami. The mutual length proportion of the head and thoracic segments can be estimated to be $12 + 4 + 3 + 2 + 2$. The head is therefore fully as long as the thorax. The length proportion of the urosomal segments is almost $1 + 3 + 2$, and the urosome constitutes not quite $\frac{1}{4}$ the total length. Here the rostrum still shows distinctly. The antennae exceed the the body's length; yet it seems only with the last segment, which, however, is comparatively very long; not counting the shaft I have been able to see sixteen segments in the antennae. In this form I have found the animal's length to vary between 0.8 and 2.1 mm.[9]

4. In the fourth stage going in reverse the larva has three thoracic segments, only two pairs of feet (the last thoracic segment is without feet) and a urosome, consisting of two segments not counting the caudal rami; but the form in general is still like that of the adult animals, with such small minor modifications, as stated in the following.

The size of this stage is from nearly 0.7 mm up to 1 mm. The mutual length proportion of the head and the three thoracic segments can almost be estimated as follows: $10 + 3 + 2\frac{1}{2} + 2\frac{1}{4}$. The head is therefore appreciably longer than the thorax. The urosome constitutes almost $\frac{1}{4}$ the total length, and the mutual length proportion of the segments is $1 + 3 + 2$. Also the head, as seen here from the side, is slightly more pointed anteriorly, or, if you will, somewhat less rounded, than in the aforementioned forms; the rostrum as yet does not seem to protrude; in the smaller individuals the first antennae are comparatively shorter, in that they generally constitute only a little more than $2/3$ the total length, whereas their size is relative to the animal's size, until it is equivalent to or even surpasses the total length;[10] their form is more robust than that of the older animals, the segments number only ten to twelve (the shaft is not included); in the small individuals the last segment is not longer than the preceding [segments], and the large lateral setae on the next to the last and the third last segment are still found to be undeveloped, which is contrary to the case in the larger individuals (of about 1 mm). The second urosomal segment is much thicker and more robust than that of the larvae of the aforementioned stages.

5. In this fifth stage the larvae are of an altogether different form, from the last mentioned, and I should not have dared give it a place here, if I had not caught it together with the remaining forms, if it had not shown itself as an obvious developmental step, and finally if a detailed examination had not shown that it belongs to the same genus as the others. The length at the most amounts to 0.5 mm. The form is rather thick, wide, robust, somewhat bowed, or apparently hunched. The greatest thickness or height constitutes nearly half its length. The head constitutes more than half the animal's length. The first antennae are short (shorter than the head) but very thick and robust, club-shaped, the shaft is nearly the same length as the flagellum; this last is not yet divided into segments, but consists of only a single, elongate-eggshaped piece, armed with many setae. The second antennae and the mandible palp are already well developed, whereas the second maxillae and maxillipeds are small. The thorax, which is much thinner than the head, consists of three segments, of which the two front ones each carry a rudimentary pair of feet. The urosome is entirely rudimentary, or especially small and in-

---

[9] In the smallest individuals of this form I have at least in part found the first antennae to be somewhat shorter than the total length, consisting of only eleven segments, yet without a trace of injury and with little change in the form. I do not know how I can explain this phenomenon.

[10] I have found this to be so in an individual of 0.95 mm length.

distinct (not half as long as the last thoracic segment), consisting of only one segment and a pair of small, seta-armed knobs or caudal rami.

6. In a still earlier stage (the earliest I know) the animal is only about a length of 0.35 mm. It deviates, as far as I have been able to observe, from the aforementioned stage only in that the separation of the segments is less distinct, the caudal rami are even more rudimentary, only two thoracic segments are found, and the thoracic feet are simply not yet developed.

I have referred all these developmental steps to the form *C. hyperboreus,* and that no doubt seems to be confirmed, partly by the proportion and the characteristics of the urosomal segments, and partly by the size of the last developmental step, which reaches 5.3 mm in length, while the adult *C. spitsbergensis* hardly reaches 4.2 mm according to my experiences. Yet I must draw attention to the fact that this case on that account cannot be considered closed, when we first remember the general law for crustacean development, namely that the species-differences do not appear in the earlier stages (often not even do the generic differences); the next to be considered is the not insignificant abovementioned size differences among forms, belonging to the same stage. It is therefore very probable that the young from several species are considered under one which, however, can be regarded as unimportant with respect to generic development in general.

I have only negative observations to report in regard to the reproductive behavior of this genus. Among the multitude of individuals I have had the opportunity to examine, I have never found any which, with an outer egg sac or egg in the body cavity, would be denoted as a female; nor have I ever found any which, with a changed form of the first antennae or one of the pairs of feet, would be denoted as a male. There rests therefore as yet just as much darkness over the sexual characteristics as over the female's method of propagation.

On the dorsal side of the cephalothorax, almost in the middle of its length, I have observed a pair of granular or cell-like bodies. I do not dare at this moment to have any opinion about whether these have any connection with reproduction, or maybe rather with digestion. The intestinal tract in the live animals has a constant wave-forming movement that reaches directly to the anus. The heart, which is shaped like an elongate sac, reaches through a large part of the cephalothorax, nearly from its front end, and makes itself known by an uninterrupted pulsation. Remarkably there is a comparatively large quantity of fat or oil, which these animals contain, and by pressing

under a glass plate it streams out to all sides;[11] therefore it remains understandable that they, in spite of their insignificant size, could furnish potent food for whales.

In the open sea they seem to hold themselves close to the water's surface, and without a doubt are frequently found in the company of the genera *Mysis, Thysanopoda,* and *Themisto* among crustaceans, with *Clio* and *Limacina* among the mollusks etc. One cannot without pleasure watch their graceful movements in a glass of water; they frequently stretch their long first antennae straight out to both sides, and keep themselves suspended with their help, and almost use them as a balancing bar. The second antennae and the mouth parts are in a constant and rapid movement, without any forward progress of the animal taking place, and therefore these movements occur for respiration and feeding.

As small as these animals are, they still possess peculiar parasitic animals, which naturally are yet much smaller. On *C. spitsbergensis* I have often found eggs fastened to the mouth parts, by what I believe to be a parasitic animal. I have also found fastened to the backs as well as to the ventral sides etc., of these crustaceans, comparatively large, irregular masses or sacs, which also seem to be egg masses of a parasitic animal. Lastly I have seen only once on a developmental step of *C. hyperboreus* (?) one small leech (of almost 1 mm length) fastened to the ventral side of the thoracic segments.

(1849) "Contribution to Carcinology (Continuation)." *Naturhistorisk Tidsskrift,* series 2, 2, 6: pp. 561–609, pl. 6.

1. *Pontia Pattersonii* Tmplt.[1]
[*Anomalocera patersonii* Templeton, 1837]
[pl. 6 (figs. 1–7) and 1845, pl. 42 (fig. 1)]

This species seems to be fairly frequent in the open sea between the Faroe Islands, Iceland, and the

---

[11] When one accepts as a rule that crustaceans show signs of fat development only to a small degree, the genus *Calanus,* then, shows a definite exception to that rule. It is also conceivable that one could in certain places and at the right time make economic use of these animals since they are so rich in oil.

[1] Robert Templeton (1837) has described and figured this animal under the name *Anomalocera Pattersonii* (*Trans. Entom. Society* [London] 2: pp. 34–40, pl. V) after specimens caught near northern Ireland, yet not entirely satisfactorily. He seems to have known little or nothing about the genus *Pontia,* founded several years earlier by Milne-Edwards, whose description of that in the third part of *Hist. d. Crustacés* (1840) disclosed clues to knowledge about the genus *Anomalocera.* [*Pontia* was first described by Milne-Edwards in 1828, the basis here for the priority which is only implied by Krøyer.]

southern tip of Greenland. Several times Captain Holbøll has brought specimens from his repeated trips between Greenland and Denmark.[2] I have caught a few specimens in northern Kattegat near Hirsholmen together with individuals of *Calanus*; others in the Atlantic Ocean off the coast of Portugal. The length of adult animals is almost up to 4.2 mm.

Both from Holbøll's and my experience, the color is no doubt the usual for pontiids, namely pretty and of a deep sky-blue with a large, white metallic or shiny mother-of-pearl spot on the dorsal surface.[3] On the other hand, Templeton states after Patterson that the color is light green with dark shadows.

The form in general is almost the same, as in the genera *Cyclops, Cyclopsina, Calanus,* etc. The largest height, which starts at the back of the head, constitutes not quite 1/5 the total length.

The head constitutes almost 1/3 the total length, and is almost the same length as the urosome, but is somewhat shorter than the thorax (the mutual proportion is almost as 7 to 9). It consists of two very distinctly separated segments that mutually have almost the same length along the dorsal side; whereas the second, on account of the oblique cutoff, is much longer along the ventral side. The anterior one of these has the eyes, the two pairs of antennae, and the labrum, while the other mouth parts belong to the second segment, which is pointed in the front and triangular or just like a pyramid, in that the dorsal surface nearly forms angles with the two lateral surfaces. When these segments are observed from the dorsal side, it is noticed, that from the first segment toward the back in the middle extends a wide, rounded-off tip, that presses into the second segment, or is taken up into a convenient indentation of its anterior edge. On both sides against the posterior edge the first segment is armed with a fairly large, posteriorly directed spine, as also its sides underneath are slightly expanded and wing-shaped; on the dorsal surface are seen several grooves and cross stripes, almost as if there were more segments.

Protruding straight down from the anterior edge of the forehead, the rostrum consists of an elongate-rectangular plate, which is cleft at the end like a fork into two strong points or spines, which are nearly as long as the plate and seem to be connected to it by an articulation. They form an angle with the base, and the points diverge weakly. They appear to be asymmetrical in form, and neither are they entirely the same in length.

And beside the base of the first antenna, almost on the border of the first segment's forehead surface and lateral surfaces, on each side, are two circular, yellow crystalline lenses, one right behind the other, and of a different size (the anterior one of almost 0.06 mm diameter, and the posterior one of 0.08 mm diameter). But beside these four lenses a far larger one is seen in the middle line of the body, although on the under surface, namely all the way forward in front of the labrum, but so close behind the rostrum, that it seems to be taken up by its rami. Particularly this last one in live or fresh animals is abundantly surrounded by a purplish red pigment.

The length of the first antennae of the female amounts to almost half its total length. The shape is seta-like or gradually tapered. One can distinguish between a 3-segmented shaft and a 22-segmented flagellum. The shaft, whose length is contained almost six times in the flagellum's length, or seven times in that of the entire antenna's length, is yet very indistinctly separated from the flagellum, and also its segments can be distinguished only with difficulty; the first is very short; the second shorter than the third and slightly pinched at the base; the third is cylindrical. The first eight segments of the flagellum are particularly indistinct, extremely short, and of far larger width than length. The following segments are plainly separated and the length dimension increases gradually, in part to a very important degree; the fourteenth to the seventeenth segments are the longest, nearly the same mutual length; the last five segments become at once considerably shorter than those just passed, but are mutually almost the same length. All the segments of the flagellum are supplied with setae on the anterior side, particularly at the end; yet they also have several setae in the middle; only one pair of the last segments also has a seta on the posterior side; the terminal segment carries a bunch of setae at the end. The setae are only of average size and of general shape.

The first antennae of the male are slightly longer than those of the female, so that one can recognize that they exceed half the animal's total length; the left one has almost the same structure as in the female; on the contrary the flagellum on the right is developed in a peculiar manner, and I therefore restrict myself to describing this one. I have not observed for certain more than nineteen segments in this; namely the first four are very indistinct, short, wide, though gradually reduced a little in thickness, yet they blend somewhat into one with the shaft, and with difficulty are differentiated from it; the next six are yet shorter and more indistinct, but also have much thinner segments, and because of the thinness they are shaped like a stem for the three following, to a swollen segment like a thick, oval knot. The fourteenth segment is of an elongate shape, thicker at the base than at the end, as

---

[2] One of these vials is found marked that the specimens contained therein were caught between 58–60° North and 5–14° West of Greenwich. On another vial is noted only that the individuals were taken in the Atlantic Ocean east of the Faroe Islands.

[3] However, pontiids lose their original color very rapidly in alcohol.

a truncated cone; the fifteenth segment is even more elongate and thin, cylindrical; likewise the sixteenth segment is very much elongate and not only strongly pinched at the base, but also acutely bent, which makes it appear as if it consisted of two segments, and which also bends the point of the antenna forward and up against the previous segments (eleventh to fifteenth), and they form a hook in connection with these. Finally the last three segments are shorter and noticeably thinner than the two just previous, but mutually of almost the same length. The anterior edge of the flagellum is armed with some hairs and setae, and also the eleventh segment has a long, bent hooklike spine at the end; and furthermore the fourteenth, fifteenth, and the first half of the sixteenth segments have a large number of especially small and tightly placed serrations, whereby without a doubt the recently mentioned hook makes a strong holding tool. In the smaller, not yet fully developed, males I have been able to distinguish twenty-two segments in the very robust flagellum, of which the first thirteen have a much greater width than length, the fourteenth has a slightly greater length than width, the fifteenth and sixteenth are very long (the longest of the segments of the flagellum), the sixteenth has a very long and strong spine at the end, which is curved slightly toward the inside, and reaches out over the two following segments, which are fairly short, while to the contrary the nineteenth to the twenty-first are long; the last segment is nearly rudimentary.

The second antennae lie much closer to the first antennae than in the genus *Calanus*, and are contained almost six times in the total length. Though on the whole of uniform construction, still they show many modifications in the individual. In the first place the basal part is nearly rudimentary (its length is comparable to the length of the anterior ramus almost as 2 to 15), and makes just one segment, whose width is much greater than its length. The anterior ramus consists of three distinct segments, whose length proportion can be estimated as $5 + 7 + 3\frac{1}{2}$. The last of these, that spreads out strongly toward the end, is not only heart-shaped but cut nearly cleft to the base into two side-tips, of which the posterior one is somewhat longer than the anterior one; this last one is armed with eight setae, for the most part of considerable length; the posterior one shows one small offset at the end, like a rudimentary segment, from which extend six very long setae. The posterior ramus is not only much thinner than that of the anterior but also much shorter, barely half the length, three-segmented, and supplied at the end with one rudimentary, three- or four-segmented flagellum; the segments' length proportion is almost $1 + 1 + 4$; and the flagellum at the end is armed with several setae of average length.

The labrum, which lies enclosed between the base of the second antennae, consists of three plates: two elongate, egg-shaped lateral plates, between which appears a little rounded-off plate; the free edges or protruding end tips of these plates are supplied with fine hairs.

The mandible, whose length constitutes almost 1/10 the total length, is distinguished by a large knot or hump, that extends almost from the middle of its anterior edge, close inside the palp's support. The fan-shaped outspread chewing surface as viewed from the side shows seven protuberances or denticles, of which the five middle ones are double; the anterior tooth is the largest, separated from the others by a much larger space than that which mutually separates them, and at the end is supplied with a spine; the two posterior protuberances are so thin that they are nearly a transition between the tooth-form and the spine-form; the last is of significant length, with serrations along the anterior edge; those in between are strong, conical. The palp, which is of the same length as the mandible, has a very small and indistinct basal segment; next a very large and wide oval-shaped segment, from whose end two small swimming rami extend; the inner of these is two-segmented, the segments are nearly the same length, at the end wide and cutoff, the first one at the end of its inner edge is armed with two long setae, the second with six setae on the wide cutoff end edge. The outer swimming ramus is slightly longer than the inner, somewhat pointed, seemingly four-segmented (yet very indistinct), and armed with five long setae at the end.

The first maxillae are very small, and, on account of their being so close together or joined together, it would have been very difficult to interpret the individual parts, if I had not had the help of the analogy of the genus *Calanus*.

The first jaw-pair consists of two plates, of which the inner—the real jaw [gnathobase]—is very short and wide and rounded off, at the end and along the inner edge [it] is armed with a dozen long spines that for the most part are located in pairs in two rows, and are supplied with small setae along the sides. The outer plate, or palp, is slightly longer and narrower, also rounded-off at the end, but supplied only with three long and strong seta-armed spines. Close up to the palp's outer edge lies a still much smaller plate, which has nearly the same shape as the palp, and likewise at the end is armed with a pair of setae. I am uncertain how I should interpret this, whether it belongs to this one or to the second jaw-pair.

The second jaw-pair is directed almost straight out from the sides, and lies very close to the outside and under the first, which is directed forward or slightly inward. [The outer lobes of the first maxilla] are first a short, wide jaw-part, slightly conical at the end, indistinctly three- or four-segmented, [and] armed on the inner (or anterior) edge and at the end with several long setae; next a shorter and narrower, but more evenly wide palp, cut off obliquely at the end, which does not show any trace of segmentation, and which is armed with four long setae; close up to this palp, and partly hidden by it, there seems still to lie a little plate armed with two setae. Finally a little farther back I observed a very small, crescent-shaped, bent out or protruding gill, armed with eight large, feather-shaped setae.

The underlip, which lies so close up to the first maxillae that it is observed and separated from these only with difficulty, consists of two bent out tips deeply separated in the middle.

The second maxillae, whose length almost amounts to 1/8 the total length, have a very strong and robust construction, and make a very strong swimming-tool, whose ramus is directed forward. The segmentation is particularly indistinct; yet I believe I can state that there are five segments just as in the genus *Calanus*. The length proportion of the segments can almost be estimated: $5 + 2 + 3 + 1 + \frac{1}{2}$. From the anterior side of the segments, with the exception of the first and maybe the last segment, knots protrude, on which swimming setae are located. I can distinguish three or four (yet very small) knots on the second segment;[4] the third segment has two large, the fourth segment one large, the fifth segment maybe one very small [knot]. The swimming-setae, of which the total count is close to twenty, are for the most part very long, strong, feather-shaped, yet the lateral setae are short and fairly sparse. It is the swimming-setae that give these maxillae their peculiar appearance.

The maxillipeds are much thinner than the second maxillae, and appear to be much smaller, though in reality they are slightly longer (very slightly more than 1/7 the total length), after the setae are deducted. This consists, just as in *Calanus,* of a two-segmented basal part and a five-segmented ramus; but on the contrary herewith it is conspicuous that the second segment of the basal part does not considerably exceed the flagellum's segment in length nor in thickness, and therefore according to appearances it could easily be regarded as a part of this. The length proportion of the basal parts to the ramus is almost as 8 to $6\frac{1}{2}$; the first segment of the basal parts to the second is as 5 to 3; the segments of the flagellum are mutually as $5 + 4 + 2 + 1\frac{1}{2} + 1$. The first segment of the basal parts is strongly swollen toward the end, and toward the front sends out three blunt, seta-bearing tips or knots, whereas the second segment does not show any like that. Even though the anterior edge of this appendage is armed with plumose setae, it has yet an entirely different appearance from the second

---

[4] Maybe rather only two knots, but each is cleft at the end like a fork, and from that extends a very long and strong plumose seta.

maxilla, in that only those [setae] on the first segment of the basal parts show considerable size and strength, and on the contrary, the remaining ones are small.

The thorax constitutes almost 1/3 the total length. The segments are very distinct, and even when viewed from the animal's side show a small protruding spine on the posterior edge of the dorsal surface; [the segments] diminish in length from anterior to posterior, yet the second and third are mutually almost the same length and together are much longer than the first. The little last segment is drawn out posteriorly on each side to a fairly small spine (a condition that seems characteristic of the whole genus). In the male this spine seems, as far as the nearest related species is concerned, as a rule to be longer than in the female.

The length of the first pair of thoracic feet amounts to 1/6 the total length or half of the length of the thorax. They are of strong construction, made of a robust, two-segmented basal part and two three-segmented rami. The basal part is slightly shorter than the outer ramus (the proportion is almost as 7 to 8), but longer than the inner (the proportion is almost as 7 to $5\frac{1}{2}$), its two segments are almost the same length; a fairly large seta extends from the inner side of the first segment. The proportion of the three segments of the outer ramus is almost $3 + 2 + 3$; each of the first two segments has a strong spine at the end of the outer edge, a long seta at the end of the inner; the last segment has two spines on the outer edge, a very long external serration at the end and four long setae along the inner edge. The inner ramus, which almost reaches to the end of the outer's second segment, has segments of nearly the same length, or the last only insignificantly longer than the two previous; a long seta extends from the inner side of the first segment, two from the second, and the third is surrounded by six long and strong setae. The setae are plumose, but the lateral setae in spite of being many in number and close together are difficult to observe because of their extraordinary fineness.

The second pair of feet is considerably longer than the first. The proportion of the basal part to the outer ramus is as 3 to 5, to the inner as 3 to 2; its first segment is recognizably longer than the second, which shows a little spine toward the end of the outer edge. The proportion of the lengths of the three segments of the outer rami is as $2 + 2 + 3$. The elongate third segment has three large spines along the outer edge and also three very small ones. The inner ramus reaches not quite to the end of the second segment of the outer ramus, and consists of only two almost equal-length segments; the first of these has three long plumose setae along the inner edge. The number and proportion of spines and setae are just like those of the first pair of feet.

The third pair of feet is like the second in both size

and shape. Neither does the fourth pair show any essential difference from the two previous pairs.

The fifth pair of feet however appears not only to be very different from the previous, but mutually is also very different in the two sexes. In the female the length is almost the same as the first pair of feet, but it is of a thin, almost delicate shape; the basal part consists of two segments, the outer ramus of two, the inner of one. The length proportion between the basal part and the outer ramus is almost as 7 to 11, and between the basal part and the inner as 7 to 2. The inner ramus is therefore nearly rudimentary; at the end it appears forklike, cleft into two large spines. The proportion of the two segments of the outer ramus, the terminal spines not included, is as 9 to 2; on the outer side the first linear segment has two large spines, one almost in the middle,[5] the second at the end; on the inner side however only one [spine] at the end, but of such considerable length that it reaches beyond, or to the end of, the spine of the second segment. The little second segment bears three strong spines at the end, of which the innermost is much longer than the other two and along its innermost edge is armed with about ten serrations. In the male this pair of feet is somewhat larger, of very robust construction, of much different shape from that described for the female; the left foot is both larger and of an entirely different shape from the right one. The right foot is in proportion in length to the left as 5 to 6; the mutual proportion of its three segments can almost be stated as: $5 + 10 + 3$; from the end of the outer edge of the elongate second segment protrudes a little spine, and one much smaller still from the middle of the third segment; which also at the end is armed with three fairly large and strong spines (especially the innermost); at the base of the third segment on the inner side is seen a little knot having several tips covered with fine hairs. The more robust left foot, whose mutual length proportion of the four segments is almost expressed by the numbers: $8 + 6 + 4 + 5$, has a large clasper, made up of the third and fourth segments. The third segment is much shorter than the previous, of almost the same width and length, at the end toward the inside it is supplied with a fairly deep notch in the edge, where a little thumb-knot, with rounded-off end, arises. The fourth segment, or the movable finger, is long and curved, but robust, only slightly pointed, supplied toward the end of its inner edge with a pair of small setae. In the young, immature male both feet are alike and of the same shape, and are somewhat more like the right foot of the adult male: they are robust,

pointed, and consist of a two-segmented basal part, whose segments are mutually almost the same length, and of a three-segmented ramus, which almost has the same length as the basal part; on the outer edge toward the end a small spine is seen on each of the segments of the ramus, and two curved spines are also found at the end of the third segment of the ramus.

In this genus the urosome seems nearly always to be of a somewhat asymmetrical shape, and shows differences in the sexes. As a rule it consists of four to five segments and the two caudal rami, but in the female all the segments are not distinct; also the urosome in the female is somewhat shorter than that in the male, which particularly depends on the caudal rami. In the female of closely related species I have found the length proportion of the parts almost as: $12 + 1\frac{1}{2} + 6\frac{1}{2} + 5 + 8$. The first segment is thick, swollen, somewhat asymmetrical; the second segment is somewhat indistinct or rudimentary, the first stands out plainly by strong pressing, several times wider than it is long; the third segment of rectangular outline, with a slightly greater length than width; the fourth segment has a slightly greater width than length, with the posterior edge bent at an angle in the middle. A fifth segment is lacking. The caudal rami are elongate, posteriorly wide and obliquely cut off but asymmetrical, or mutually somewhat uneven; also in this respect they differ; whereas the right one is supplied toward the back with four very long plumose setae and on the outside of these with a strong spine, on the contrary the left one is armed with five long plumose setae on the edge farthest back; both have very fine hairs on a part of the inner edge and a strong seta at the end of this. A much elongate, saclike spermatophore is often found fastened to the first segment of the urosome. In the male I have found the urosome to consist of five segments (therefore one more than in the female, and all distinct) without the two caudal rami. The proportion of the parts is almost: $6 + 5 + 6 + 4 + 3 + 10$. The first segment spreads out posteriorly on the right side, in the shape of a large spine or as a horn. The caudal rami are of a particularly elongate and narrow shape, posteriorly slightly wider, each supplied on the posterior edge with five very long plumose setae (the longest more than twice as long as the caudal rami), and along the inner edge with very fine hairs.

2. *Pontia Edwardsii* Kr. n. sp.
[*Labidocera acutifrons* (Dana, 1849)]
[pl. 6 (figs. 8–11)]

I have found a few specimens of this form in the Atlantic Ocean, close to the South American coast, at 21° South, in the month of June.

The color is a beautiful indigo-blue with a large, silver-colored shield on the dorsal surface of the first

---

[5] It is possible that an articulation is found at the place where this spine is located, and that therefore the ramus is three-segmented, but the segmentation in this case is especially indistinct.

three thoracic segments and with purple-colored eyes.[6]

This form is fairly thick and robust, with a strongly swollen and arched dorsal surface. The largest of the individuals I have had occasion to examine had a length close to 6.4 mm; the thickness constituted nearly a third of the length, and the height was not much less than the thickness.

The head's first segment, whose length constitutes almost a sixth the total length, has a somewhat conical shape, in that its dorsal surface slopes downward very much, and the sides converge forward; the dorsal surface is wide and fairly flatly rounded-off, the posterior lateral-angles are sharp and slightly protruding. The forehead is armed with one perpendicular downward-pointing rostrum, which is almost half as long as the segment, plate-shaped or strongly squeezed together in its anterior-posterior direction, nearly the same width in its whole length, cut out at the end and the tips of each side made up thereby armed with a fairly long, very thin, straight spine.

The eyes are three in number, very small (their diameter is seldom 1/12 the length of the first segment), circular, purple-colored,[7] two located above and slightly to the front of the first antennae on the side of the head, the third on the under surface slightly behind the rostrum.

The first antennae, which are fastened on the under surface of the head, close behind and under the eyes, constitute slightly more than 2/3 the total length, and are setiform. The shaft, as usual, is very small—it constitutes almost 1/7 the length of the antennae— [and] three-segmented. In the flagellum of the female I have counted twenty-two segments, of which the first seven are particularly short; those that follow gradually increase in length, yet in this way, that the thirteenth to the seventeenth are the longest, and on the contrary the last five again decrease. On the inner side of the flagellum is observed a multitude of setae, of which those placed toward the base are fairly long and for the most part are plumose setae.

The second antennae are much shorter than the first, and are contained almost six times in the total length, but their construction is fairly strong. From an especially short foundation extend two branches or rami, of which the posterior one is very delicate and also much shorter than the anterior; the proportion of their length is almost as seven to twelve. The posterior ramus shows only one quite distinct segment, or bears at the end an extraordinarily small, entirely rudimentary segment. The anterior ramus

consists of four segments, though in part slightly indistinct; the second of these is the longest; the third consists of almost 1/5 the length of the ramus, but is strong, at the end somewhat expanded; the fourth is very small, nearly rudimentary. The setae, with which both rami are armed at the end, have an extraordinary length, as most of them somewhat exceed the length of the ramus, from which they originate; most of them by far are plumose setae; on the posterior ramus I have observed six, on the anterior twelve, of which half of them extend from the rudimentary fourth segment.

The mandible is of strong construction, quite hornlike, fan-shaped and expanded at the end and cleft into seven large, very pointed, and also somewhat curved teeth or spines. Among the five lowest of these spines are located some especially small setae, and under the seventh is a single, large and strong seta. The palp consists of a very short foundation segment pinched at the base, from which extends one large, somewhat oval plate, which is lengthened or drawn-out at the end on the outer side; [the plate] bears two small two-segmented branches, which at the end are supplied with some extremely long plumose setae (five on each branch).

The construction of the first maxillae is nearest to that in the previous species. The underlip consists of two plates, united in the back, on the end edges and the inner edges armed with short, somewhat backward-curved setae or spines.

The second maxillae are fairly small, in that their length, not counting the setae, constitutes only almost 1/7 the total length. They consist of an oval basal part, a five-segmented ramus, and five very large spines, pointing forward, extending from the anterior side of the basal parts toward the end. The length proportion of the ramus to the basal parts is as five to four; its first two segments are long, the last three very short; on the first side several spines extend from the segments, those along the one side are closely armed with spinules or serrations; the spine extending from the base of the first segment has teeth along its inner edge, on the contrary the other spines have teeth along their outer edge; and these teeth are comparatively longer. The five spines extending from the basal parts seem to have too great a length to be called spines, but on the other hand are too strong to be named setae; the last two of these, which are the longest, are almost twice the length of the ramus; also these are armed along the one side (the side pointed toward the body) with smaller spines, that form two parallel rows in their length; these spinules are partly too thin, and sit too far apart from each other to form a saw, but seem rather to produce a comb.

The maxillipeds, not counting the setae, are fairly short (their length is only almost 1/6 the total length)

---

[6] The silver-colored shield usually maintains itself fairly well at least for a time in alcohol; while on the contrary the other colors change to a dirty gray-brown or black-brown.

[7] The three lenses are yellow, but the whole space between them fills up with a purplish-red pigment mass.

and of a robust construction. As well as the indistinctness of the segmentation has allowed me to judge, they have the same proportions as in the previous species.

The first pair of feet, when the setae are not counted, is contained seven times in the total length, and is therefore short, but of strong construction. The foot consists of a short, two-segmented basal part and two three-segmented swimming rami, of which the outer is much longer than the basal part, the inner slightly shorter. The segments of the outer swimming rami, of which the first and the last are mutually almost the same length and much longer than the second, are all armed at the end of the outer edge with a very large and strong spine (the third [segment] also has one in the middle); there are also some small spinules at the base of the spines; this swimming ramus is supplied on the inner side with six long setae, one extending from the second segment, the others from the third segment; the first five are plumose setae, the sixth and outer one is closely armed with serrations along the outer edge. The inner swimming ramus is armed with eight long plumose setae, one on the first, one on the second, and the others on the third segment, which also bears a strong spine at the end. The two segments of the basal part are each supplied with a large plumose seta on the inner side at the end.

The second pair of feet is somewhat longer than the first (its length amounts to slightly more than 1/5 the total length) but is almost of the same construction and shape; only the number of setae and the length proportion show some small modifications.

The third pair of feet is again slightly longer than the second (its length amounts to slightly more than 1/4 the total length), and is essentially of the same construction, yet the shape is somewhat more elongate; the outer swimming ramus seems to have five segments, the inner two. The number of spines on the outer edge increases here: the second segment of the basal parts has one, the first segment of the outer swimming ramus three spines, the second segment three, and each of the following three two spines.

The fourth pair of feet agrees in nearly all respects with the third pair. Yet it can be noticed, that there are some plumose setae extending from the inner side of the first segment of the basal part, which on the first pair of feet are distinguished by their size and strength, and which on the following pairs gradually decrease, here becoming nearly rudimentary or imperceptible.

The fifth pair of feet in the female is shorter than the three previous (its length amounts to almost 1/6 the total length) and is essentially different in construction, though it also consists of a two-segmented basal part and two rami. It does not have any plumose setae, and it appears to have a jumping-form instead of a swimming-form. It is incurved like a

crescent, the basal part is of the same length as the outer end-branch, its second segment is somewhat longer than the first (their length proportion is almost as two to three). The outer, longest end-branch seems to consist of two segments (yet indistinct); it is armed on the outer side with two small but robust spines, and at the end it appears cleft into three branches or three strong spines, of which the outermost is very small, the middle one very large, the innermost only slightly smaller than the middle one. It has one spine still much larger and stronger, than any of those previously mentioned, which is almost in the middle of the inner side. The inner branch, which is almost the same length as the second segment of the basal part, and which slightly exceeds half the length of the outer branch, seems almost to consist of two indistinct segments, of which the first is somewhat shorter than the second; this last is cleft at the end like a fork into two robust but pointed branches, which differ somewhat in length.—In the male this pair of feet is of robust and strong construction; the right foot, which greatly exceeds the left in size, forms a robust clasper, and consists of four segments, which mutually have nearly the same length. The first is conical; the second is bent out like a crescent on the outer side, on the inner armed with a pair of setae; the third segment or hand is very swollen, of much greater width than length, with a crescent-shaped notch toward the front; at the end on the inner side it is supplied with a fairly large and pointed thumb-knot and a seta; the robust, just barely pointed but angularly curved fourth segment, or finger, has a little setose spine toward the end. The left foot, which is hardly more than half as long as the right, shows a very indistinct segmentation, excluding only the last segment; the first two segments are almost the same length; the third and last, which is contained seven times in the length of the foot, is nearly circular [and] armed on the inner side with fine, closely set hairs and at the end with three spines.

The thorax consists of five segments, of which the first is the longest (just as long as the second segment of the head), the following two are the same length, together slightly longer than the first, the fourth is almost half as long as the third, and the fifth still shorter than the fourth.[8] The fifth segment is so closely wedged to the fourth, that together they seem to constitute only one; under a strong magnifying glass one first discovers the line of segmentation between them; in the female the sides of the fifth segment lengthen posteriorly into two oval, nearly setiform elongate bodies, pointed at the end; in the male on the contrary only into a small spine.

The urosome is short and in the female is also

---

[8] All these lengths mentioned here are, as is customary, calculated along the dorsal middle line.

robust or wide, consisting of four segments, not counting the caudal rami, of which the first two segments are much closer together, or shaped like cones which fit into each other, and in the mid-dorsal surface both are drawn out posteriorly to a point, which lies out over the third segment. The second segment is supplied with a pair of small, hornlike, slightly incurved jumping legs, which end with a spine. The third segment is posteriorly cut off straight, and from its mid-posterior edge a somewhat smaller fourth segment extends, which posteriorly appears to be deeply indented, nearly cut in two; to the two tips of these segments are fastened two somewhat oval plates, the caudal rami. Each of these is armed on the outer and posterior edge with five long, somewhat curved plumose setae, and also with a small one, which extends in direction inward and posteriorly from the inner posterior corner.

In the male the urosome is of a much more elongate and narrow form, fairly symmetrical, consisting of five distinct segments not counting the caudal rami. The length proportion of these parts is almost $4 + 3 + 3 + 2\frac{1}{2} + 2\frac{1}{2} + 7$. The first segment is recognizably wider than the following; but all show a rectangular outline with a larger width than length; it must be remarked about the last, that its posterior edge is acutely bent in the middle. The caudal rami are of elongate-oval shape, and are armed posteriorly with six plumose setae, of which the innermost is short, the others much longer (even longer than the entire urosome). The inner edge of the caudal rami is supplied with fine close-set hairs.

3. *Pontia Nerii* Kr.[9] [new species]
[*Labidocera nerii* (Krøyer, 1849)]
[pl. 6 (figs. 12–16)]

Only a pair of specimens of this species was caught in the Atlantic Ocean, out from Cape Finisterre, on the ninth of October, 1840.

The color is light-blue, shading somewhat toward greenish. The size is almost 3.2 mm.

The rostrum is large, in that when the basal part is included, it is contained just ten times in the total length. The branches are longer than the basal part (the proportion to that is almost as three to two), nearly completely parallel and straight, very thin and pointed, without a sign of a segment or an articulation with the basal part, very widely separated from each other (their mutual separation is nearly the same as their length). Perhaps a rudimentary eye lies between the branches by this base.

Near the back of the rostrum on the head's upper surface or crown appear two small, circular points, the eyes, which are mutually widely separated, or

closer to the lateral edges of the head. The lenses are almost spherical, and have a very dark yellow-brown color and a fairly distinct hardness; the pigment, which is a purplish-red, is so abundant that it covers the whole forehead- or crown-surface anteriorly, and in such a way in the live animals that it seems to combine both lenses together as one eye. The pigment disappears entirely in alcohol.

The first antennae are of average length (almost the same as 4/5 the total length), and distinguish themselves with a number of strong plumose setae, which are located anteriorly on the shaft as well as on the first five or six segments of the flagellum. The other segments of the flagellum have fairly large and stiff plain setae; some located on the posterior side of the last three segments particularly distinguish themselves by their length. The right antenna of the male does not seem to swell up as much as that on the closely related species.

The second pair of antennae is not remarkable, except perhaps that the short posterior ramus is fairly robust.

The postero-lateral spines on the head's first segment seem to be entirely missing, or are very small and pressed close to the side of the head, because I have not been able to observe them with certainty.

I have not found anything to discuss about the mouth parts.

I have observed only four thoracic segments, whose length proportion can almost be expressed with the numbers $4 + 3 + 3 + 3$. The fifth thoracic segment is entirely hidden under the fourth, and it is this last one that sends out a spine on each side of the posterior edge of the thorax; [the spines], however, in both sexes are only of an insignificant size.

The characteristics of the feet are: (1) that the two-segmented basal part is very short, insignificantly longer than the inner ramus, but hardly more than half as long as the outer; (2) that the inner ramus is not even half as long as the outer, consists only of two segments, and bears nine long swimming-setae on the last one of these on the first pair of feet, seven on the following three; (3) that the last segment of the outer ramus has an extremely deep indentation on the outer edge and along both sides is extraordinarily robust, with wide edges of integument supplied with spines, and also a very large, sword-shaped, finely serrate terminal spine.

The right foot of the fifth pair of feet in the male, or the forceps, consists of three segments, whose mutual length proportion can almost be estimated as: $5 + 8 + 10$; the basal parts are comparatively fairly thin; the second segment, or hand, is strongly curved or crescent-shaped on the anterior side, distinguished by an especially large and pointed spine (at least compared to the length of the foundation segment), extending from the base of the posterior edge at a

---

[9] Named in this way after the place where it was found: *Nerium promontorium*, Finisterre.

ght angle; the third segment, or claw, is distinctly
)nger than the hand, but very thin; the whole length
 nearly the same thickness, strongly curved, is not
)inted at the end but is bluntly rounded off; it bears
 small spine on the inner edge toward the base and
vo small curved, or apparently hook-shaped, setae
: the end. The left foot is somewhat shorter, also
·ree-segmented; the proportion of the segments is
'most: 4 + 3 + 5; the first segment is slightly bowed;
·e second fairly straight and cylindrical; the third
·gment somewhat curved, lance-shaped, or with the
·se strongly expanded and rounded-off, very strong
·d gradually pointed at the end, the posterior edge
· the first half supplied with a fine fur or with short,
·rly hairs.—The fifth pair of feet in the female is
·ry small, and consists of a short, one-segmented (?)
·sal part, a long, one-segmented outer ramus or
·ther stylus, which has a linear form, and at the end
 cleft forklike into three spines or points, of which
·e middle one considerably exceeds the length of
·e two others; like a rudiment of the inner ramus a
·nall oval plate must be noticed, which lies at the base
· the stylus on the inner side, and is contained three
· four times, or more, in its length.

The urosome, which in the male constitutes slightly
·ore than 1/5 the total length, and is thin, consists of
·ve segments (of which the first is nearly entirely
·idden by the last thoracic segment) and two caudal
·mi, whose mutual length proportion can almost be
·esignated by the numbers: 4 + 3 + 2 + 1 + ½ + 5.
·he form is symmetrical; the caudal rami are short,
·ide, cut off obliquely, with six setae at the end (a
·ry short inner one and five long ones, of which the
·ngest are just as long as the urosome, or slightly
·nger). In the female the urosome has nearly the
·me length as in the male, but it is thicker, of a
·)mewhat asymmetrical form, and consists only of
·vo segments and caudal rami; the mutual proportion
· these parts is almost 6 + 1½ + 4.

The smallest of my specimens (of almost 2.1 mm
·ngth) is the same as the male in the construction
· the urosome, on the contrary as the female in the
·iape of the antennae. I must assume that this is a
·oung male, in which the designated reproductive
·nplements have not yet reached their development.

[4.]　*Ifionyx typicus* Kr. [1845]
[*Candacia pachydactyla* (Dana, 1849)]
[pl. 6 (figs. 17–21) and 1845, pl. 42 (fig. 3)]

I have caught some individuals of this new form in
·ie Atlantic Ocean near the Equator (4° North)
·round mid-July. I have also found them at 30°
·outh in the month of May.

The size seldom exceeds 3.2 mm.

The color, I assume, must have been like the usual
·ky- or indigo-blue of the closely related forms, as I

have not noted anything about that. The specimens
preserved in alcohol appear to be reddish yellow.

Its greatest height is not quite contained four times
in the total length.

The head constitutes slightly more than 1/3 the total
length, and is somewhat shorter than the thorax. It
does not show any sign of separation into two segments, nor any spines on its sides. The form is
somewhat pointed anteriorly.

The rostrum, which is hidden between the first
antennae (not located, as in the previous genera, ahead
of these), is especially small, not cleft but simply
pointed.

I have not definitely located the eyes or crystalline
lenses, neither on the upper surface of the head nor
on the under surface behind the rostrum.

The first antennae, which are located at the end of
the head's anterior point (not, as in the genera
Calanus and *Pontia*, a distance behind the anterior
edge of the head), are almost as long as the animal's
total length, or are only very slightly different from
that. They consist of a cylindrical, 3-segmented
shaft, whose segmentation is yet very indistinct, and
a 21-segmented flagellum. The shaft is contained
almost 6½ times in the length of the flagellum; its
third segment is the largest, and is distinguished by a
protruding swelling, which bears a bunch of setae
or spines, [and] which is almost in the middle of the
anterior side. The segments of the flagellum show
some irregularities, in that the shorter and the longer
segments are mixed up among themselves without
order; some of the first segments are also fairly indistinctly set apart. The fourteenth to the sixteenth
segments are the longest; those following decrease,
so that the nineteenth is very short, whereas the last
two increase considerably. The segments of the
anterior edge are in part armed with long and strong
setae, but for a still larger part [are armed] with
large spines; the last segment has at the end a large
tuft of setae, which spreads itself radiantly to all
sides.

The second antennae are contained six times or
slightly more in the total length; they consist of a short,
two-segmented shaft and two flagella, which have
nearly the same length, but the posterior one is much
thinner than the anterior. The anterior flagellum is distinguished by the fact that the short, heart-shaped
second segment has two quite distinct, although rudimentary, small segments at the end, placed side by side
(one on each of the tips of the heart). The posterior
flagellum first shows a very short segment, which is
elongate or linear, and at the end two rudimentary
segments. The length proportion and number of
swimming setae of both flagella are almost as in the
previous genera.

The labrum consists of two crescent-shaped plates,

deeply separated from each other in the middle, sparingly supplied on the edges with fine setae.

The mandible, which seems to have a segmentation separating it into an outer and an inner piece, is of a very thin and narrow, slightly S-shaped curved form; at the inner end it is cleft forklike into two pointed and thin branches or spines, but does not possess any denticles; on the posterior edge of the posterior branch is a small tuft of very fine hairs. The palp appears in this respect diverted, in that its inner ramus is not placed at the end of the large segment, by the side of the outer, but on the contrary is entirely on the inner side of the segment, close to its base; the first of the two segments of this ramus is extremely small, whereas the two small-segments of the outer ramus have almost the same length. The basal part, by which the large, disk-shaped segment is fastened to the mandible, shows here a much stronger development than in *Calanus* or *Pontia,* and seems to consist of two segments. The large segment has a somewhat elongate, rectangular form.

The first maxillae are represented here by a form differing somewhat from that described above in *Pontia* and *Calanus.*

The first jaw-pair consists of a large, somewhat rectangular basal segment, from which a jaw-plate and a palp extend from the inner half of the end edge.[10] The jaw-plate is comparatively very small, slightly pointed, armed along the inner edge with about ten spines placed in a double row, of which [the spine] located at the end is remarkable by its extraordinary strength. The palp is more than triple the length of the jaw-plate, but yet is extremely narrow, linear, weakly bent, and armed at the end with three setae, of which the middle is short, the two others long.

The second jaw-pair, which is much shorter than the first, when the base part is included, and in itself is somewhat shorter, when the base segment is not included, consists of a large oval jaw-plate, which at the end produces a very small two-segmented palp, [and] on the outer side at the base a similarly very small, but comparatively strong protruding gill. The jaw-plate is wide, bluntly rounded-off, supplied on the inner side with only three medium-sized setae, [and] which only shows very weak signs of lateral-setae; two longer setae extend from the end. The palp is almost fork-shaped in that its first segment is much wider than its second, with its inner angle strongly protruding; each of its segments is armed with two setae; those on the last segment have an extraordinary length, but the setae seem not to possess lateral-setae. Instead of a crescent shape the gill

here seems to have a short conical shape, and is armed with at least seven in part long plumose setae.

An underlip of a fairly complicated construction, made of several rounded-off, hirsute tips, is present, but I am not in a position to state positively how many tips there are; I have definitely observed four, but I suppose that there are more.

I now come to the parts, that by their extraordinary size and strength most conspicuously distinguish the nearest genus, namely the second maxillae. The length amounts to almost 2/5 the total length. They seem to consist of only three segments, whose mutual length proportion can almost be set as: $35 + 17 + 3$; its whole length has a considerable thickness; extending from the anterior side of the base of the first segment are four setae placed in pairs, toward the end is a pair of very large and strong spines; on the anterior side of the second segment are two still much larger spines (longer than the segment itself); finally there is the small, fairly indistinct third segment, or the rudiment of several segments fused into a flagellum, armed with three spines or hooks of surprising size and strength.

The maxillipeds by comparison can be called rudimentary; they are contained more than five times in the total length, and are much shorter than the first segment of the second maxillae; the construction however is fairly strong; they consist of a two-segmented shaft and a flagellum. The length proportion of these parts is $1 + 4 + 4$. The first segment does not have any setae on the anterior edge; the second segment has three short plumose setae in addition to some especially short hairs. The robust, seven-segmented flagellum shows about ten medium-sized plumose setae.

The thorax, which slightly exceeds the head in length, consists of five segments, whose length proportion can almost be set as $3 + 3 + 3 + 2 + 2$. On the posterior sides the last thoracic segment is drawn out into a spine just as in *Pontia;* yet it is only of insignificant size.

The first four pairs of feet, which have an average size and a very strong construction, correspond in general to the proportions in the genera *Calanus* and *Pontia;* from the first to the third they increase gradually in length, whereas the fourth again decreases, or has almost the same length as the second; they consist of a two-segmented basal part and of two rami. The basal part does not have quite half the length of the outer ramus, or is almost like its first two segments, but exceeds the inner ramus in length; its two segments have nearly the same size; a plumose seta extends from the inner side of the end of the first [segment]. The outer ramus consists of three segments, of which the first two are almost the same length, the third longer than both of the previous two together; a certain characteristic must be noticed, that the outer edge of the whole ramus is more or

---

[10] The basal segment however may be considered as common to both jaw-pairs, in that the second extends from the outer half of the end edge.

less finely serrate.[11] It is the third, elongate-oval segment of the ramus, which has the most distinct and strongest saw-weapon (on the last two-thirds), and that segment is also distinguished, by its considerable length and width, and by its brown or nearly black horn-color (wherewith the animal is recognizable at once, without closer examination) and by its apparently leaflike construction; extending right from the base all the way to the point there is a longitudinal nerve (actually, as it seems, a slightly protruding keel), that parts the segment into two lateral halves, and from the longitudinal nerve on the outer side two side nerves extend, on the inner [side] five [extend], that reach all the way to the edge. Five plumose setae are placed along the inner edge of this segment; one extends from each of the previous two segments. The inner ramus is still more rudimentary than in *Pontia* and *Calanus,* and consists only of two segments, of nearly the same length, which are supplied with a dozen long plumose setae.

The fifth pair of feet is very small, nearly rudimentary (in length almost the same as 1/9 the animal's total length), consisting of a two-segmented basal part and a ramus, in that the inner ramus here completely disappears. The construction is comparatively very strong. The basal part is almost the same length as the ramus; its first segment is somewhat longer than the second, club-shaped; the second segment barrel-shaped. The ramus consists only of one segment, which in its whole length has almost the same width, is weakly bent inward; on the middle of the outer side it is armed with a large spine-like protrusion, and at the end cleft into three very large and strong, pointed, horn-colored spines, of which the middle one is the largest; between this one and the innermost a roundish, flat knot can be observed. Three spinulose setae are located fairly close together on the inner side of the ramus toward the end.

The urosome, which constitutes somewhat more than 1/4, but not fully 1/3, the total length, consists of four segments and two caudal rami; the length proportion is almost: $6 + 4 + 5 + 2 + 3$. The first three segments are mutually very distinctly separated from each other and from the fourth, in that those that follow are narrower than the foregoing; the fourth on the contrary is fairly indistinctly offset from the caudal rami. Moreover, the urosome's form is symmetrical. The characteristics of this are: that its first segment appears to be supplied with a pair of limbs, which are in length proportion to the urosome as 3 to 10; these limbs are spiniform, and consist of two segments or divisions, of which the first is swollen or somewhat blister-like, the second elongate and conical. The caudal rami also seem to be two-seg-

mented, though very indistinct; each is armed with five long plumose setae (somewhat shorter than the urosome), of which one extends from the first segment of the ramus, the others from the second. The shape of the caudal rami is slightly pointed.

[5.] *Centropages typicus* Kr. [new
genus and species]
[pl. 6 (figs. 22–26)]

A few specimens of this form, nearly all males, were caught in the Atlantic Ocean, a few miles from Cape Finisterre, almost in mid-September.

The length of the adult animals seems hardly to exceed 1.6 mm.

The shape in general is as in the genus *Pontia,* yet perhaps not quite as high. The greatest height is contained almost five times in the total length, the greatest thickness almost four.

The length of the head amounts to 1/3 the total length or slightly more. Its two segments, measured along the dorsal surface, appear to be almost the same length. The shape of the head, when viewed from the side, is fairly strongly pointed anteriorly. On the contrary when viewed from above, the lateral edges appear to be somewhat wave-shaped, and the anterior part of the head to be abruptly squeezed in, and limited to four straight lines, from which three angles arise.

The rostrum is divided into two small, pointed branches, but more of its condition I am unable to describe.

I have observed with certainty only one crystalline lens (of almost 0.04 mm diameter), lying between the base of the first antennae.

The first antennae, which in length are equal to the total length, or even slightly exceed it, consist of a 2-segmented shaft and a 22-segmented flagellum. The first eight to ten segments of the flagellum are fairly short, those that follow gradually become particularly longer; the last five are somewhat shorter than the nearest previous, but still very elongate and mutually almost of the same length. The shape of the segments is nearly linear (not club-shaped). The setal armature is fairly sparse, as far as the first half of the flagellum is concerned, and seems to be entirely missing from the last half. On the last segment of the shaft and on the third segment of the flagellum I have always observed a strong, anteriorly-curved spine. It is like this in the female. In the male the proportion of the shaft and the first ten segments of the flagellum is like that in the female, but the following five segments (eleventh to fifteenth) swell up slightly, so that together they have a very elongate spindle shape; the sixteenth is again cylindrical and very elongate, the seventeenth is still longer than the sixteenth, but also noticeably thinner; the last five segments are shorter, of general form, or

---

[11] In addition there are four or five large, movable spines, which also occur in the genera *Pontia* and *Calanus.*

almost like those in the female. A large spine appears on the anterior edge of the fourteenth segment, [on] the fifteenth segment a comb with very fine teeth; also on the sixteenth segment and almost on the first half of the seventeenth are seen numerous very fine serrations or cross-stripes. A strong spindle-shaped muscle stretches from the base of the eleventh to the end of the sixteenth segment, and serves well by its drawing these segments together to form a grasping tool.

The second antennae are small (they are contained almost six times in the total length) but of strong, even robust, construction and with both rami equally developed. The two-segmented basal part is somewhat shorter than the rami; the anterior ramus, as usual, is three-segmented, the first segment the largest. In the posterior ramus I have been able to distinguish six segments: first a large and robust one, the next three very small, somewhat pinched-in, the next again a very large one and finally, as usual, a rudimentary terminal segment.

I have not been able to describe the labrum.

The mandible is expanded at the inner end, ear-shaped, armed with seven small teeth. The outer ramus of the palp is somewhat conical, four-segmented, armed with five robust setae. The inner ramus is two-segmented, the first segment with a protruding tip or expansion on the inner side; both segments are supplied with long setae.

I have examined well the first maxillae of this little animal, but have not been entirely able to describe them in detail, so that I feel it is inappropriate to give a description of them.

The second maxillae, whose length is contained almost five times in the total length, almost match in form with the maxillipeds in the genus *Calanus*.

The maxillipeds are distinctly longer than the second maxillae (their length exceeds 1/4 the total length) but are of the same shape.

Five distinctly protruding thoracic segments are present, and the fifth exceeds the nearest previous [segment] in length, even when its long lateral spines are not counted. The length proportion of the segments can almost be expressed by the numbers: 4 + 2 + 2 + 2 + 3. The posteriorly directed lateral-spines of the fifth segment in the female are longer, more pointed, and thinner than in the male, and the posterior edge of the segment protrudes slightly sinuate, or shows signs of a pair of outbendings.

The thoracic feet are fairly large and strong (their length is contained nearly four times in the total length), the first and last pairs are slightly shorter than the three middle pairs, which are mutually almost of the same length. The two-segmented basal part is almost half as long as the outer ramus, but is almost of the same length as the inner; both rami are three-segmented, the outer with five spines on the outer side

and at the end with a long, wide, flat, saberlike spine, which is very finely serrate along the outer edge. The inner ramus is of very thin and weak construction when compared to the outer.

The fifth pair of feet, which, as mentioned, is somewhat shorter than the previous three, is separated in the female from the rest of these only by a very large, dagger- or spurlike spine, which extends on the inner side from the end of the second segment of the outer ramus (in a posterior direction, but also slanted somewhat inward), and almost reaches to the end of the third segment of the ramus (the terminal spine is not counted). On the contrary in the male this pair of feet has a very different and peculiar shape, which shows quite distinctly with what simple modification the forceps arise. The right foot, only with which is the forceps concerned, is very large, larger than any of the other feet. It consists of a two-segmented basal part, with almost the usual construction, and of two three-segmented rami, of which the inner has a shape almost like that in the preceding pair of feet. But on the outer ramus the first segment is short, slanted, and almost rectangular, nearly of the same width as length; at the end of the outer edge it is supplied with a small spine; the second segment is rounded off at the base, very strongly expanded or swollen, at the end of the inner edge drawn out to an extremely large and strong, slightly curved claw, which represents the thumb of the forceps, while the rounded-off basal part forms the hand. The strong, somewhat elongate and conical third segment with its very long terminal spine, which together form a bent bow toward the thumb, represents thereby a two-segmented movable finger; a small spine is observed toward the end of the inner edge of the third segment. In the male the left foot consists of a two-segmented basal part and two three-segmented rami, but still with a characteristically different appearance from the previous pairs of feet, because the outer ramus here is small (almost not noticeably longer than the inner or basal part) and armed only with quite small spines, or without such a long, sword-shaped spine at the end, which the other feet possess.

The urosome, which amounts to almost a third the total length, has in the female a somewhat asymmetrical shape, and consists, in addition to the caudal rami, of three distinct segments; the mutual length proportion of these parts can almost be expressed by the numbers 3 + 5 + 2 + 3. A fairly large, slightly curved spine, with a few small lateral setae on it, extends from the middle of the posterior edge of the dorsal surface of the first, somewhat thick and robust segment, [and] two long and thin, needle-shaped spines [extend] from the ventral surface. The caudal rami are of elongate, linear shape, slanted and bluntly rounded-off posteriorly, each armed with four long setae, the outermost and the innermost also with a spine. In the male the shape of the urosome is more elongate, thin, and regular; one can distinctly recognize four segments not counting the caudal rami, whereas the first segment is small; the mutual length proportion can almost be expressed by the numbers: $1 + 3 + 3 + 2 + 3\frac{1}{2}$. No spines are observed on any of the segments, as they are in the female.

[6.] *Agetus typicus* Kr. [new genus and species]
[*Corycaeus* (*Agetus*) *typicus* (Krøyer, 1849)]
[pl. 6 (figs. 27–29)]

I have caught only a single specimen of this anomalous form in the Atlantic Ocean at almost 43° North in mid-September.

The color is white, with a delicate reddish hue.

The size amounts to not quite 1.6 mm.

The head is of fairly large size, particularly when compared with the four thoracic segments; its length nearly constitutes half the total length (or it amounts to a little more than 3/7 the total length), and is more than three times as large as the length of the thorax, but only insignificantly exceeds the urosome. When observed from above, the form of the head appears as an elongate oval, [which] posteriorly gradually narrows slightly, which unites with the thorax in a straight cut-off line, and which anteriorly is very flat or blunt. From the side it has a more noticeable form, in that only its dorsal surface is very weakly, or nearly not at all, arched, the anterior end is blunt or cut off straight, while on the contrary the under surface is delineated by two straight lines, which almost in the middle, where the mouth parts have their place, come together to form a sharp angle, which is nearly a right angle, or at least not very blunt. The greatest height of the head is therefore recognizably larger than its width or thickness and greater than one-fourth the animal's total length. A rostrum was not observed.

Most noticeable in this form is the extraordinary size of the eyes at the anterior end of the head; they include nearly entirely the cut-off or very bluntly rounded-off surface of the forehead, and are mutually separated only by an insignificant space. The eye's diameter amounts to nearly 1/8 the total length. The purplish-red pigment has disappeared in this animal preserved in alcohol, and therefore one observes only two amber, circular, watchglasslike concave lenses.

On the under surface of the head, under and between the eyes, extending from a little knot or a protruding beam, are the short first antennae, which are contained seven times or perhaps slightly more in the total length. They consist, as well as I have been able to observe without dissecting them, of six fairly robust segments, and are threadlike; a division into a shaft and flagellum seems not to have taken place, or, in other words, the segments are not very different in length and thickness. Several small setae are located along the lower edge of the antennae and at the end.

Close behind the first pair of antennae protrude the second; they not only are distinguished by their considerable size (which is almost the same as 1/3 the total length) but also by their peculiar form, which very strongly reminds one of the form of the grasping feet in the genus *Squilla* [Stomatopoda]. They seem to consist of only four segments: two short and also somewhat indistinctly offset basal segments, a very large and anteriorly elongate slightly bulging hand, and a thin, curved claw, which has almost the same length as the hand and in its whole length has almost the same thickness. A long spine or a very large seta, which crosses the claw, extends from the near-mid anterior edge of the hand.

I have nothing to report in detail about the mouth parts, as I do not want to destroy the only individual of this strange form; [the mouth parts] are located far behind the second pair of antennae, almost at the middle of the under surface of the head, right where it makes a protruding angle; and they protrude even more strongly than this [angle], and appear as a large blunted knob, having nearly the appearance of a suction tool. On each side at the posterior base of the knob I have noticed a small, straight posteriorly directed spine. Between the mouth parts and the feet is a large open space, which does not seem to be occupied by jaw-feet; which however only results from the fact that the jaw-feet are so close to the other mouth parts. By pressing I have observed altogether four pairs of mouth parts; the first three I assume to be mandibles and two pairs of maxillae; the last, which considerably exceeds the previous in length, and apparently forms a kind of grasping tool, or as a form which comes somewhat close to the second antennae, becomes also the maxillipeds.

The thoracic section is very short, in that it almost amounts to only 1/7 the total length; it seems however to consist of five segments, though it no doubt must be admitted that these are very indistinctly marked. A very large and strong spine extends from each side of the next to the last [segment]. The length proportion of the thoracic segments seems almost to be expressed by the numbers: $2 + 4 + 3 + 3 + 3$. Posteriorly the thoracic section decreases considerably in thickness, so that its shape resembles an inverted, blunted bowling pin.

Five pairs of thoracic feet are present, which decrease in length anterior to posterior, yet like this, so that the first three pairs mutually show only a little difference in length; the last two on the contrary are much shorter than the previous. All the feet are swimming feet of ordinary construction, consisting of a basal part and two rami, which on the innermost or posterior edge are supplied with long setae; the outermost rami are also [supplied] with spines on the anterior edge and at the end. This concerns only the first three pairs; each of the last two small [pairs] seems to consist of only two small segments, but I cannot definitely report their condition, for by pressing they have become invisible, or covered by the larger feet.

The urosome, which constitutes 3/7 the total length, shows, when it is seen from the side, an elongate and thin form, and seems to consist of two segments in addition to the caudal rami (the length proportion of these parts can be described by the numbers 3 + 1 + 2). On the contrary when observed from above the relation appears somewhat different, or three segments in addition to the caudal rami seem to be present, in that the first wide segment, posteriorly supplied with a notch, spreads itself like a pair of wings out over the second, so that only a very small part of it shows from the back. The caudal rami are long, entirely linear, each supplied at the end with a fairly strong but not very long spine (almost 1/3 the length of the caudal rami).

[7.] *Thaumaleus typicus* Kr. [1845]
[as *Thaumatoessa typica* Krøyer, 1845]
[pl. 6 (figs. 30–31) and 1845, pl. 42 (fig. 4)]

In the following lines, I shall try to present an introduction of a highly remarkable, even in some respects entirely enigmatic, new form, of which I have examined only a single individual. I caught this in March, 1839, at Bejan (inlet to Trondheimfjord), close to the beach, at ebbtide among the seaweeds. The animal's movements were quick and powerful; the integument leatherlike and unusually soft for a crustacean (nearly as in a worm).

The length amounted to slightly over 4.2 mm.

The color was very unusual for a crustacean, namely onion-green with cinnabar-red antennae and feet. The eyes, if these organs are present, are not differentiated in color from the main color of the body.

The form is somewhat elongate (nearly five times as long as high or thick), fairly plump, anteriorly blunt, gradually sharply pointed posteriorly. One can plainly distinguish a large head, which consists of only one piece, five thoracic segments, and a small thin urosome.

The head, when it is measured along the ventral surface or along the sides underneath, is just as long or even slightly longer than the thorax, and amounts to only a little less than half the total length; on the contrary its posterior edge in the middle of the dorsal surface shows a very deep indentation or apparently a notch, which almost includes half the length of the head, and which is filled up by the extending, tongue-like prolongation of the anterior edge of the first thoracic segment; if the measurement of the notch is taken into consideration (which I did not do in the table of measurements), the thorax becomes much

longer than the head. The head is blunt and rounded off anteriorly, and on the under surface somewhat swollen.

Only one pair of antennae, which is only of average length (it is almost the same as 1/4 the total length), but of very robust construction, extends from the anterior surface of the head. The robustness makes it very difficult to determine the border between the shaft and the flagellum; however, I believe that I can include the first four segments with those. Their mutual length proportion is almost 5 + 3 + 3 + 3, and together they compare themselves to the flagellum almost as 14 to 11, or are slightly longer than it; yet this distribution is fairly arbitrary. The flagellum consists of five segments, of which the first exceeds the others so much both in length and thickness, that it might better be included with the shaft; namely it is longer than the flagellum's four other segments together; a large and strong spine or spiniform seta extends from the end of the last segment. The whole flagellum as well as the last two segments of the shaft are supplied with so many large setae, which in this way are so twisted together that it becomes impossible to determine their number, or from which segment each extends; these setae are of an entirely uncommon thickness or coarseness, but are also unusually soft, and show in a peculiar manner a knotted and uneven or apparently grainy upper surface.

Just as with the second pair of antennae, the head shows not the least sign of mouthparts. Its undersurface is entirely closed, if I may say so, or the dorsal shield continues without the slightest interruption on the undersurface of the head.

The length proportion of the thoracic segments, if no attention is given to the first segment's above-mentioned tongue-shaped protrusion, [which is] enclosed in the notch of the head, can almost be described by the numbers 5 + 4 + 3 + 3 + 4. Each one of these segments is supplied with a pair of feet of average length, but also, as far as the four anterior [feet] are concerned, are of a very strong construction. The first pair of feet, [when] straight posteriorly and not including the setae, reaches almost to the end of the third thoracic segment, the second pair somewhat beyond the fourth, the third pair slightly beyond the fifth, the fourth pair to the end of the urosome, and the fifth pair slightly beyond that. The construction of the feet is in general like that in the genus *Pontia*.

The length of the first pair of feet amounts to nearly 1/4 the total length; its wide, two-segmented basal part is longer than the rami (its proportion to those is almost as 3 to 2); its first segment is nearly three times as long as the second; the segmentation is fairly indistinct. The three-segmented rami are nearly the same length, or the inner only slightly shorter than the outer; this last with only a single large spine on the outer side (extending from the end of the first segment); five or six [12] long and extremely robust setae extend from the last segment of each of the rami, and one [such seta] from the inner side of each of the previous segments; these setae are not plumose setae, but of a construction like those located on the antennae. The last segment of the rami is the largest, the second the smallest.

The second pair of feet seems to me to differ from the first, only in that the second segment of the basal piece is even smaller in proportion to the first, and that a spine extends from the end of the first segment on the inner side.

The two following pairs of feet do not remarkably differ from the second.

On the contrary the fifth pair of feet is somewhat shorter than the previous (its length is contained almost six times in the total length), of a very thin and elongate form, and consists only of a very small, one-segmented basal part and one ramus or rather a stylus. The proportion in length of the basal parts to the stylus is almost as 1 to 6. The stylus is very thin, plump, without a sign of segmentation, sharply pointed and curved at the end, so that it forms a hook. It shows no spines or setae.

The urosome is very small (its length is contained nearly eleven times in the total length), and consists of only two segments and of the caudal rami; the mutual length proportion of these parts can be expressed by the numbers $3 + 2 + 5$. The caudal rami are widely rounded off at the end, and each supplied posteriorly with five long setae and the outermost with a small spine. The setae exceed the urosome in length (the proportion of these is almost as 3 to 2); the construction is similar to that of the setae on the antennae and feet.

The described specimen was a female, whose interior was full of a great many small eggs (of 0.04 to 0.05 mm diameter); it can be found—or rather what remained of the pieces, which could be saved after the examination—in the Royal Natural History Museum's systematic crustacean collection. The illustration is in *Voyage de Scandinavie etc., Crustacés* pl. 42 fig. 4, *a–e*.

In the following lines I report the characteristics and the diagnoses, not only of the species and new genera described above, but also of a couple of new exotic species, which I did not mention in the foregoing.

1. *Pontia Pattersonii* Tmplt.
[*Anomalocera patersonii* Templeton, 1837]

[Latin] *Form* rather elongate; height not quite equaling one-fifth length. *Rostrum* large, deeply bifurcate, rami asymmetrical, unequal, and subdivergent. Five purple *eyes,* four on the head, and the fifth below

---

[12] Six from the outer ramus, five from the inner.

and behind rostrum. *First antennae* less than 2/3 animal's length, not much thicker at base, with short, simple setae. *Second antennae* make up almost one-sixth of animal's length. Fairly large postlateral *hooks* on first segment of head. *Urosome* equals approximately one-third of the animal's length.

♀. *Fifth thoracic foot* straight and slender; outer ramus much longer than basal part, with serrated terminal inner spine of last segment not exceeding inner spine of prior segment; inner ramus rudimentary. *Urosome* consists of four segments in addition to caudal rami: first asymmetrical and bulging in middle; second subrudimentary, without feet; third square, and fourth dorsal as well as middle emarginated. *Caudal rami* asymmetrical and obovate, not quite equaling one-fourth of length of urosome.

♂. *Left fifth thoracic foot* of male subchelate and strong, consists of four segments, the third of which square, projecting in crescent in front of middle, with small, blunt, protuberant node; right foot slightly shorter than left, with three segments and barbed tip. *Urosome* rather elongate, consists of five distinct segments, the first of which enlarged on right and dorsal, and extends quite a bit as a horn; other segments symmetrical; *caudal rami* greatly elongate, sublinear, subclavate, equaling about one-third of urosome.

2. *Pontia Edwardsii* Kr. [new species]
[*Labidocera acutifrons* (Dana, 1849)]

Form rather robust, dorsally convex; height equals about one-third of length. *Rostrum* large, deeply bifurcate, with parallel and equal rami. Two very small, purple *eyes* slightly anterior to crown of head, and a solitary eye below and behind rostrum. First antennae equal about 3/4 animal's length, thickened at base, armed with plumose setae. *Hook* of first head segment very small, not conspicuous. *Second antennae* exceed one-fifth of animal's length. *Urosome* equals one-fourth of length.

♀. *Fifth thoracic foot* of female strong, robust, curved; its outer terminal ramus approximately equal to, or very slightly larger than, length of basal part, with three-forked tip, barely twice as long as two-forked inner ramus. *Urosome* consists of four segments in addition to caudal rami: first and second produced beyond middle in long, acute angle; second very short, with two minute rudimentary feet [see note, p. 42]; third segment square; fourth narrower than the rest, dorsally split very deeply (almost divided into two lateral parts).

♂. *Right fifth thoracic foot* of male subchelate, strong, consists of four segments, the third of which (the hand) much wider than long, hollowed out in crescent before the middle, with large protuberant, pointed node; left foot much shorter, scarcely half length of right, blunt, indistinctly articulated. *Urosome* rather elongate and narrow, subsymmetrical, consists of five distinct segments (none of which has feet) and elongated oval caudal rami.

I have caught specimens in the Atlantic Ocean (21° South). Length approximately 6.4 mm.

3. *Pontia Nerii* Kr. [new species]
[*Labidocera nerii* (Krøyer, 1849)]

Form rather robust; height little more than one-fourth length. *Rostrum* very large, deeply bifurcate (fork longer than basal part). Crown of head with *two* minute, separated, lateral, purple *eyes*. *First antennae* approximately equal to animal's length, armed with large, plain setae. Without distinct post-lateral *hooks* on first segment of head. Only four *thoracic segments* dorsally evident (fifth covered by the others). Inner ramus on *thoracic feet* very short, two-segmented; third segment of outer ramus with deep incisions on outer edge, strong marginal spines, and serrated, sword-shaped terminal spine.

♂. *Urosome* approximately equal to one-fifth of animal's length. *Fifth* thoracic foot on *right side* of male subchelate, three-segmented, distinguishable by very large spine on base of hand, and mobile digit longer than hand, slender, arched, and blunt. Fifth foot on left side slightly shorter, three-segmented, second segment cylindrical, straight; third segment curved in, armed with little point. *Urosome* rather slender, symmetrical, consists of five segments and short, enlarged, obliquely truncate caudal rami. Right first antenna less swollen than usual.

♀. *The fifth pair of feet* on female minute, two-segmented, styliform, trifurcate at tip, with rudimentary forward inner ramus. *Urosome* rather thick, asymmetrical, consists of two segments and caudal rami.

4. *Pontia brachyura* Kr. n. sp.
[? *Labidocera detruncata* (Dana, 1849)]

Form rather subrobust, height approximately equal to one-fourth length. *First antennae* reach approximately 3/4 animal's length, *second antennae* approximately one-fourth. In addition to *two* minute *eyes* on crown there is a conspicuous *third,* located below and behind rostrum. *Hooks* on first segment of head minute but distinct. Fork of *rostrum* very shallow, scarcely equaling one-fourth of rostrum's length. Postlateral *angles* of fifth thoracic segment lengthened into large spines. *Urosome* approximately equal to one-fifth of length, not going beyond tip of fourth thoracic foot. *Caudal rami* equal to one-third of length of urosome.

♀. Posterior margin of female's fifth thoracic segment with two large, rounded protuberances between the lateral spines. Fifth foot biramus, each ramus one-segmented; outer ramus barely longer than basal part, curved or hooked, with barbed outer edge; inner ramus very small, bifurcate. *Urosome* robust, asym-

metrical, consists of three segments in addition to caudal rami.

♂. Right *first antenna* of adult male greatly thickened in middle, subchelate. Fifth thoracic foot uni-ramus; right foot large, subchelate, subslender, with very long, laminar claw or mobile digit; *left* foot somewhat shorter, strong, three-segmented, each segment of approximately equal length. Posterior *protuberances* on margin of fifth thoracic segment not very obvious. *Urosome* symmetrical, consists of five segments and caudal rami.

I have examined specimens captured near Pulo-Pinang, East Indies [Pulau Pinang].

Ifionyx Kr.[13] [1845]
[*Candacia* Dana, 1846]

Head approximately equal to length of thorax, about 1/3 total length, and consists of one segment, armed with an extremely small non-bifurcate rostrum, situated between the first antennae. Head not armed with postlateral hooks. There seem to be no distinct *eyes.* Posterior ramus of *second antennae* approximately equal in length to anterior ramus, but much more slender. *Mandible* deeply bifurcate at tip, but not armed with any teeth; lateral palp with inner ramus. *Second maxillae* very large (almost as long as head), very strong, and armed with incurving spines of unnatural size. *Maxillipeds* quite subrudimentary, not equipped with any spines. *Thorax* composed of five distinct segments, the last of which with dorso-lateral spines. *Thoracic feet* serrated on outside of outer ramus; third segment of this ramus very large, an elongated oval, horny, blackish-brown, and similar in construction to the veins of a leaf; inner ramus minute, two-segmented. *Fifth foot* minute, with a solitary one-segmented jumping (?) ramus. *Urosome* of female consists of four segments and caudal rami; the first segment with jumping feet. [The question mark is Krøyer's. The expression used here is the Latin "pedibus saltatoriis," used also in the Latin description for *Pontia edwardsii,* but translated there as "rudimentary feet," since the same structure was referred to in figure 10,*f* legend as "pedibus rudimentariis."]

[5.] *If. typicus* Kr. [1845] [*Candacia pachydactyla* (Dana, 1849)]. Head somewhat shorter than thorax. Fourth and fifth thoracic segments each very distinct. Flagellum of maxilliped six- or seven-segmented. First thoracic foot slightly less than one-fourth of animal's length. First, second, third, and fourth thoracic feet with horny last segment on outer ramus.—I have caught specimens in the Atlantic Ocean (4° N.).

[6.] *If. orientalis* Kr. n. sp. [?*Candacia ethiopica* (Dana, 1849)]. Head longer than thorax. Fourth and fifth thoracic segments not too distinct. Flagellum

---

[13] Ἰφι, strong; ονυξ, claw.

of maxilliped four-segmented. First thoracic foot quite a bit more than one-fourth of animal's length. First, second, third, and fourth thoracic feet with completely horny outer rami.

I have examined only two specimens, captured near Pulo-Pinang, East Indies.

Centropages Kr.[14] [new genus]

Head approximately as long as thorax, one-third of animal's length. Head consists of two distinct segments, with only one *eye* between the bases of first antennae, with bifurcate *rostrum*, but without lateral hooks. Posterior ramus of *second antennae equal to anterior ramus in length and thickness*. First antennae and feeding-appendages very similar to those of *Pontia*. Five distinct *thoracic segments*; the fifth with postlateral corners stretched out into large barbs, bi-sinuate on posterior margin. The four anterior pairs of thoracic *feet* each with very long sword-shaped *blade* [15] on their *terminal outer rami*, with conspicuous serrated outer edges; the rest as in *Pontia*. *Fifth* pair of *feet* of both sexes biramus; on *female* shaped as preceding feet, except second segment of outer ramus armed with very large and strong inner blade. *Right* foot of *male* larger than the rest, with subchelate hand on outer ramus, and bi-articulate mobile digit, which is elongate, slender, and curved inward; *left* foot short, of swimming type, without blades. *Urosome* of female asymmetrical, composed of three segments and caudal rami; [urosome] of male symmetrical, [composed of] four distinct segments in addition to caudal rami; first segment of female armed posteriorly with dorsal and ventral spines; first segment of male unarmed.

[7.] *Centr. typicus* Kr. [new species]. Blade of

second segment of outer ramus of female's fifth foot rather slender, somewhat incurved, and not at all diverging from third segment, or making very acute angle with it. Length of female's urosomal segments and caudal rami indicated by the following numbers: $3 + 5 + 3 + 2$; first segment armed with solitary mid-dorsal spine on posterior edge, but two ventral spines.

[8.] *Centr. chilensis* Kr. n. sp. [*Centropages brachiatus* (Dana, 1849)]. Blade of second segment of outer ramus very strong, straight, and diverges strongly from third segment, sometimes at nearly a right angle. Length of urosomal segments and caudal rami of female approximately indicated by the following numbers: $5 + 5 + 2 + 4$; first segment armed with two dorsal and two ventrolateral spines.

I have caught a few specimens in the Pacific Ocean near the coast of Chile; length approximately 2.1 mm.

Agetus Kr.[16] [new genus]
[Corycaeus Dana, 1845]

Head much longer than thorax, consists of one segment, approximately half the animal's length, with two very large and distinct *eyes,* two pairs of *antennae,* the anterior of which short and threadlike, the posterior large and subchelate; lacking *rostrum.* *Maxillary feet* [second maxillae and maxillipeds] subchelate. *Thorax* very short, composed of five not very distinct segments, produced posteriorly into two lateral *spines,* and with five pairs of swimming *feet.* *Urosome* consists of three segments (the first of which dilated and clytraeform, the second almost completely hidden), and two elongate, styliform caudal *rami.*

Thaumaleus Kr.[17] [1845, as *Thaumatoessa*]

Head consists of one segment, large, almost equaling half the animal's length, deep dorso-posterior incision, with only one pair of very strong *antennae,* lacking *rostrum.* *Eyes* and feeding-appendages apparently lacking. *Thorax* large, composed of five distinct segments, without spines, but with five pairs of swimming *feet,* the last of which uniramus, simple, and hooked (in the female). *Urosome* very small, consists of two segments, and two sublinear caudal rami.

---

[14] Κεντροπαγης, greatly barbed.

[15] [Danish] I use the expression *stylus* [blade] here and elsewhere instead of *aculeus* [spine], not on account of a peculiar construction, but because of size and particularly thickness, which at the base is so great that it includes the whole edge of the segment; so that one can say with greater justice that the segment widens out and lengthens to this *stylus,* and that this extends out from the segment. It is also different from other spines in that its base does not articulate with the segment, but is a direct continuation of it.

[16] Αγητος, extraordinary.

[17] Θαυμαλεος, something to be marveled at.

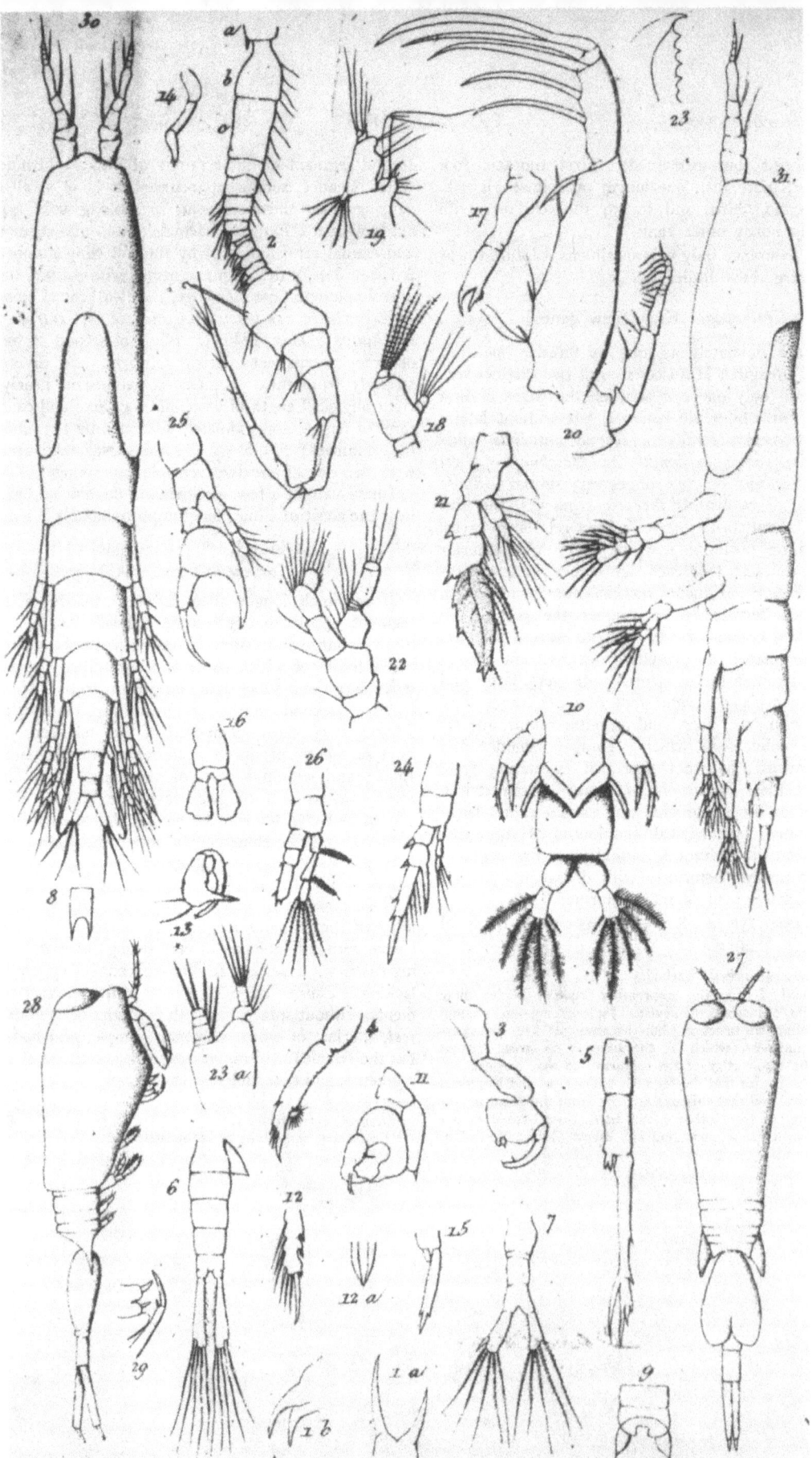

Plate 6 from "Karcinologiske Bidrag" in: *Naturhistorisk Tidsskrift*.

Explanation of Plate Six

Figs. 1-7. *Pontia Pattersonii* Tmplt.

Fig. 1,*a*. Rostrum, shown from above; *b*. same, seen from side.

Fig. 2.   Right first antenna of male; *abc*. base.

Fig. 3.   Right fifth thoracic foot of male.

Fig. 4.   Left fifth foot of male.

Fig. 5.   Fifth foot of female.

Fig. 6.   Urosome of male with caudal rami.

Fig. 7.   Urosome of female.

Figs. 8-11. *Pontia Edwardsii* Kr.

Fig. 8.   Rostrum, seen from above.

Fig. 9.   Last three thoracic segments of female, shown from dorsal surface; *a*. third segment; *b*. fourth segment; *c*. fifth segment.

Fig. 10. Urosome of female, shown from dorsal with fifth thoracic foot on right side [*sic*]; *abcd*. fifth foot; *ab*. basal part; *c*. terminal outer ramus; *d*. terminal inner ramus; *e*. first urosomal segment; *f*. second segment with rudimentary feet (*g*); *h*. third segment; *i*. fourth segment; *k*. caudal rami.

Fig. 11. Right fifth thoracic foot of male.

Figs. 12-16. *Pontia Nerii* Kr.

Fig. 12. Last segment of outer ramus of fourth thoracic foot; 12,*a*. spine of this segment, magnified.

Fig. 13. Right fifth thoracic foot of male.

Fig. 14. Left fifth foot of male.

Fig. 15. Fifth foot of female.

Fig. 16. Urosome of female (with setae of caudal rami removed).

Figs. 17-21. *Ifionyx typicus* Kr.

Fig. 17. Mandible.

Fig. 18. Mandible palp.

Fig. 19. First maxilla; *a*. basal part; *b*. anterior jaw; *c*. palp of this jaw; *d*. posterior jaw; *e*. palp of posterior jaw; *f*. gill plate.

Fig. 20. Maxillary feet; *a*. second maxilla; *b*. maxilliped.

Fig. 21. First thoracic foot.

Figs. 22-26. *Centropages typicus* Kr.

Fig. 22. Second antenna; *a*. basal part; *b*. anterior ramus; *c*. posterior ramus.

Fig. 23. Terminal part of mandible; 23,*a*. mandible palp.

Fig. 24. Fifth thoracic foot of female.

Fig. 25. Right fifth thoracic foot of male.

Fig. 26. Left fifth foot of male.

Figs. 27-29. *Agetus typicus* Kr.

Fig. 27. *Ag. typicus*, shown from dorsal surface.

Fig. 28. Same, shown from side. *a*. head; *b*. thoracic segments; *c*. urosome; *d*. eyes; *e*. first antenna; *f*. second antenna; *g*. buccal parts; *h*. thoracic feet; *i*. urosome.

Fig. 29. Buccal parts, magnified.

Figs. 30-31. *Thaumaleus typicus* Kr.

Fig. 30. *Th. typicus*, shown from dorsal surface.

Fig. 31. *Th. typicus*, from side.

Measurements [mm]	Calanus spitsbergensis Kr.	Calanus hyperboreus Kr.	Calanus minutus Kr.	Calanus affinis Kr.	Calanus quinquenulatus Kr.	Calanus cristatus Kr.	Calanus caudatus Kr.	Agetus typicus Kr.	Thaumaleus typicus Kr.
Total Length	4.2	3.9*	1.6	2.5	3.7	9.5	3.2*	1.5	4.5
Greatest Height	1.06	0.85	0.35	0.53	0.85	2.12	—	0.42	0.95
Length of Head	1.48	1.17	0.48	0.85	1.48	3.52	0.30	0.70	2.12
Rostrum	0.42	—	—	—	0.35	0.42	—	—	—
First Antenna	4.51	4.24	1.27	2.76	4.45	10.6	2.76	0.21	1.17
Shaft	0.42	0.42	—	—	0.74	0.85	—	—	0.53
Flagellum	4.03	3.82	—	—	2.86	9.75	—	—	0.64
Second Antenna	0.95	—	—	—	0.64	1.48	—	0.53	X
Mandible	0.32	—	—	—	0.35	0.70	—	—	X
Mandible Palp	0.27	—	—	—	0.35	0.70	—	—	X
First Jaw-pair	0.32	—	—	—	0.30	0.64	—	—	X
Second Jaw-pair	0.32	—	—	—	0.27	0.64	—	—	X
Second Maxilla	0.53	—	—	—	—	0.85	—	—	X
Maxilliped	0.85	0.74	0.25	—	0.64	1.91	—	—	X
First Thoracic Segment	0.74	0.58	0.21	0.42	0.42	1.70	—	0.42	0.53
Second Thoracic Segment	0.42	0.35	0.11	0.21	0.32	0.85	—	0.08	0.42
Third Thoracic Segment	0.42	0.35	0.11	0.21	0.27	0.85	—	0.06	0.32
Fourth Thoracic Segment	0.32	0.27	0.11	0.16	0.21	0.64	—	0.06	0.32
Fifth Thoracic Segment	0.27	0.27	0.05	0.11	0.27	0.42	—	0.06	0.42
First Foot	0.85	0.74	0.21	0.53	0.74	1.48	—	0.32	1.06
Second Foot	1.17	0.95	0.32	0.64	0.95	—	—	0.25	1.06
Third Foot	1.27	1.17	0.42	0.70	0.85	—	—	0.21	1.06
Fourth Foot	1.38	1.17	0.42	0.70	1.27	—	—	0.08	1.06
Fifth Foot	0.95	0.85	0.21	0.21	0.74	—	—	0.08	0.74
Urosome	1.17	0.95	0.48	0.64	0.95	1.91	1.17	0.64	0.42
Caudal Ramus excl. setae	0.32	0.21	0.08	0.14	0.21	0.32	—	0.21	0.21
Urosome's Thickness	—	0.21	0.07	0.16	0.21	0.64	—	0.21	0.21

[* Disagrees significantly with text.]

Measurements [mm]	Pontia Pattersonii Tmplt. ♂	Pontia Pattersonii Tmplt. ♀	P. Nerii Kr. ♂	P. Edwardsii Kr. ♀	Pontia brachyura Kr. ♀	Pontia brachyura Kr. ♂	Ifionyx typicus Kr. ♂	Centropages chilensis Kr.	Centropages typicus Kr.	Ifionyx orientalis Kr.
Total Length	4.0	4.4	3.1	5.9	4.2	4.0	3.2	2.1	1.6	2.8
Greatest Height	0.74	0.85	0.85	1.91	1.06	1.06	0.74	0.42	0.32	0.70
Length of Head	1.27	1.38	1.17	1.06	1.38	1.27	1.17	0.74	0.53	1.17
Rostrum	—	0.32	0.32	0.53	0.32	0.42	—	0.08	—	—
First Antenna	2.54	2.33	2.50	4.24	3.18	3.07	2.82	2.12	1.70	2.44
Shaft	0.32	0.32	0.42	0.64	0.53	0.53	0.42	0.21	—	—
Flagellum	2.23	2.01	2.12	3.82	2.65	2.54	2.44	1.91	—	—
Second Antenna	0.70	0.74	0.74	1.27	1.06	0.95	0.42	0.42	0.27	0.42
Mandible	0.42	0.42	0.21	0.53	0.42	0.42	0.35	—	—	—
Mandible Palp	0.42	0.42	0.21	0.53	0.42	0.42	0.35	—	—	—
First Jaw-pair	0.21	0.21	—	0.74	0.32	—	0.42	—	—	—
Second Jaw-pair	0.32	0.32	—	0.85	0.32	—	0.21	—	—	—
Second Maxilla excl. setae	0.53	0.53	0.42	—	0.70	0.64	1.17	0.32	0.32	1.06
Maxilliped excl. setae	0.64	0.64	0.53	1.06	0.64	—	0.51	0.64	0.42	0.42
First Head Segment	0.64	0.70	0.64	0.64	0.70	0.64	—	0.42	0.27	—
Second Head Segment	0.64	0.70	0.64	1.06	0.70	0.64	—	0.32	0.32	—
First Thoracic Segment	—	0.53	0.42	1.06	0.74	0.64	0.32	0.21	0.17	0.27
Second Thoracic Segment	—	0.42	0.32	0.64	0.42	0.32	0.32	0.16	0.08	0.27
Third Thoracic Segment	—	0.42	0.32	0.64	0.42	0.32	0.32	0.16	0.08	0.21
Fourth Thoracic Segment	—	0.32	0.32	0.32	0.42	0.64	0.21	0.16	0.08	0.11
Fifth Thoracic Segment	—	0.16	X	0.21	0.21	0.32	0.21	0.11	0.13	0.11
First Foot	0.64	0.70	0.64	0.85	0.70	0.64	0.64	0.42	0.32	0.85
Second Foot	0.85	0.95	0.85	1.27	1.17	1.06	0.95	0.53	0.42	0.95
Third Foot	0.85	0.95	0.85	1.59	1.27	1.27	1.17	0.53	0.42	1.06
Fourth Foot	0.85	0.95	0.95	1.70	1.27	1.06	0.95	0.42	0.42	0.95
Fifth Foot	0.85	0.70	0.64	1.06	0.70	0.70	0.32	0.42	0.32	0.27
Urosome	1.48	1.40	0.64	1.48	0.85	0.85	0.85	0.64	0.48	0.64
Caudal Ramus excl. setae	0.53	0.32	0.21	0.42	0.27	0.27	0.14	0.18	0.11	0.11
Urosome's Thickness	0.27	0.35	0.18	—	—	0.21	0.30	—	—	0.27

INDEX